Design of Wind and Earthquake Resistant Reinforced Concrete Buildings

W0018364

Design of Wind and Earthquake Resistant Reinforced Concrete Buildings

Somnath Ghosh and Arundeb Gupta

CRC Press
Taylor & Francis Group
Boca Raton London New York

CRC Press is an imprint of the
Taylor & Francis Group, an **informa** business

First edition published 2021
First by CRC Press
6000 Broken Sound Parkway NW, Suite 300, Boca Raton, FL 33487-2742

and by CRC Press
2 Park Square, Milton Park, Abingdon, Oxon OX14 4RN

© 2021 Taylor & Francis Group, LLC

CRC Press is an imprint of Taylor & Francis Group, an Informa business

Library of Congress Cataloging-in-Publication Data
Names: Ghosh, Somnath (Civil engineer), author. | Gupta, Arundeb, author.
Title: Design of wind and earthquake resistant reinforced concrete
buildings/Somnath Ghosh & Arundeb Gupta.
Description: Boca Raton : CRC Press, 2021. |
Includes bibliographical references and index.
Identifiers: LCCN 2020056207 | ISBN 9780367537791 (hbk) |
ISBN 9781003083320 (ebk) | ISBN 9780367537821 (pbk)
Subjects: LCSH: Earthquake resistant design. | Wind resistant design. |
Buildings, Reinforced concrete–Design and construction.
Classification: LCC TA658.44 .G53 2021 | DDC 693.8/5–dc23
LC record available at https://lccn.loc.gov/2020056207

ISBN 13: 978-0-367-53779-1 (hbk)
ISBN 13: 978-0-367-53782-1 (pbk)
ISBN 13: 978-1-00-308332-0 (ebk)

Typeset in Times
by Newgen Publishing UK

Dedicated to our parents, wives and children

Contents

List of Figures xi
List of Tables xv
Preface xix
Acknowledgements xxi
Authors xxiii
Notation xxv

Chapter 1 Introduction 1

    1.1   Preamble 1
    1.2   A Few Important Aspects of Structural Design 1
            1.2.1   Strength and Serviceability 2
            1.2.2   Ductility and Hysteresis 2
            1.2.3   Redundancy 4
    1.3   Architectural Requirements 4
    1.4   Lateral Load-Resisting System 9
            1.4.1   Subsystems and Components 9
            1.4.2   Moment-Resisting Frames, Braced Frames
                      and Shear Walls 10
    1.5   Collapse Pattern 12
    1.6   Dynamic Response Concept 14
    1.7   Wind Load and Earthquake Load 16
            1.7.1   Wind Load 16
            1.7.2   Earthquake Load 17

Chapter 2 Wind Analysis of Buildings 19

    2.1   Preamble 19
    2.2   Wind Load Provisions as per IS 875 (Part 3), 2015 21
            2.2.1   Different Approaches to Wind Analysis 22
                      2.2.1.1 Pressure Coefficient Approach 23
                      2.2.1.2 Drag Coefficient Approach 24
                      2.2.1.3 Gust Factor Approach 25

Chapter 3 Seismic Analysis of Buildings 29

    3.1   Preamble 29
    3.2   Seismicity 29
    3.3   General Principles and Design Criteria 32
    3.4   Response Spectrum of a Ground Motion 34
            3.4.1   Acceleration Response Spectrum of a
                      Ground Motion 34
            3.4.2   Liquefaction Potential 34

	3.5	Estimation of Base Shear	49
		3.5.1 Various Aspects of Base Shear	49
		3.5.2 Estimation of Base Shear as per IS 1893 (Part 1), 2016	52
		3.5.2.1 Equivalent Static Method	52
		3.5.2.2 Response Spectrum Method	54
	3.6	P-Δ Analysis	55
	3.7	Ductility Assessment	55
	3.8	Reinforced Concrete Buildings with Unreinforced Masonry Infill Walls	56

Chapter 4 Structural Design of Reinforced Concrete Buildings ... 57

	4.1	Preamble	57
		4.1.1 Steps for Structural Design of Reinforced Concrete Framed Buildings	57
	4.2	List of Relevant IS Codes	58
	4.3	Load Calculation	58
		4.3.1 Dead Load	58
	4.4	Design Example of a Six-Storied Reinforced Concrete Framed Residential Building	59
		4.4.1 Choice of Beam Depth	60
		4.4.2 Choice of Slab Thickness	61
		4.4.3 Calculation of Dead Load	62
		4.4.4 Live/Imposed Load	64
		4.4.5 Approximate Axial Load on a Particular Column	64
		4.4.6 Design of Slab Panels	66
		4.4.7 Wind Load Analysis	71
		4.4.7.1 Basic Wind Pressure	71
		4.4.7.2 Wind Load as per "Drag Coefficient Approach"	73
		4.4.7.3 Wind Load as per the "Pressure Coefficient Approach"	78
		4.4.8 Seismic Load Analysis	81
		4.4.9 Substitute Frame Analysis under Dead and Live Loads	89
		4.4.10 Frame Analysis under Wind and Seismic Forces	116
		4.4.11 Summary on Maximum Bending Moment and Shear due to Dead Load, Live Load, Wind Load and Seismic Load	131
		4.4.12 Design of Frame Beams	146
		4.4.13 Design of Columns	151
		4.4.14 Design of Foundations	165
		4.4.15 Working Drawings of Slabs, Beams, Columns and Foundations	168

4.5   Design of a 15-Storied Reinforced Concrete-Framed
        Residential Building on a Pile Foundation 175
        4.5.1   Dead Load and Live Loads....................................... 175
        4.5.2   Wind Analysis 177
                 4.5.2.1   Basic Wind Pressure................................ 177
                 4.5.2.2   Wind Load as per "Drag Coefficient
                           Approach" 180
                 4.5.2.3   Wind Load as per "Pressure
                           Coefficient Method"................................ 181
                 4.5.2.4   Wind Load as per "Gust Factor
                           Approach" 184
                 4.5.2.5   Wind Load Analysis Using Software 195
        4.5.3   Seismic Load Analysis Using Software 195
        4.5.4   Different Checks.. 195
        4.5.5   Design of Beams, Columns and Pile Caps Using
                 Software.. 197
        4.5.6   Working Drawings of Slabs, Beams,
                 Columns and Foundations 197

Chapter 5   Comparison of Basic Parameters Stipulated for Wind and
            Seismic Analysis, as per IS, IBC, ASCE, ACI, EN and
            BS Codes 207

            5.1   Preamble 207
            5.2   Wind Load Analysis... 209
            5.3   Seismic Load Analysis....................................... 212
            5.4   Numerical Example of Wind and Seismic Load
                    Analysis.. 213
            5.5   Comparison of Basic Parameters Stipulated in Indian,
                    American and British Codes 221

Bibliography 227
Index... 231

Figures

Figure 1.1   Re-entrant corner 6
Figure 1.2   Slab panels with cutouts 7
Figure 1.3   Out-of-plan offsets 7
Figure 1.4   A non-parallel system 8
Figure 1.5   Strength irregularity 14
Figure 1.6   Mass irregularity 15
Figure 3.1   Reverse faulting 30
Figure 3.2   Normal faulting 31
Figure 3.3   A seismograph 32
Figure 4.1   Plan of a six-storied building 59
Figure 4.2   Sectional elevation of a six-storied building 60
Figure 4.3   Influence area for the column marked B4 64
Figure 4.4   Marking of slab panels 66
Figure 4.5   Frontal area for joint A1 against the wind load 74
Figure 4.6   Influence area for joint 2b against the wind load 74
Figure 4.7   Building frame in the shorter direction (4–4) 75
Figure 4.8   Building frame in the longer direction (B–B) 76
Figure 4.9   Plan showing the wind angle and external wall faces 79
Figure 4.10 Wind loads on frame 4–4 (along the shorter direction of the
                 building) 82
Figure 4.11 Wind loads on frame B–B (along the longer direction of the
                 building) 82
Figure 4.12 Frame in the shorter direction showing the seismic load
                 in each joint 88
Figure 4.13 Frame in the longer direction showing the seismic load
                 in each joint 88
Figure 4.14 Typical floor load dispersion for dead and live loads 89
Figure 4.15 Substitute frame A1–B1–C1–D1 at first floor level in the shorter
                 direction (4–4) 89
Figure 4.16 Dispersion of floor loads on frame 4–4 91
Figure 4.17 Load transferred from floor to substitute frame in the shorter
                 direction (Figure 4.14) for getting maximum support
                 moment at B 91
Figure 4.18 Free and elastic shear: (a) span AB and (b) span BC 95
Figure 4.19 (a) Loading diagram for getting maximum support at B1,
                 (b) bending moment diagram and (c) shear force diagram of the
                 substitute frame in the shorter direction (Figure 4.15) 96
Figure 4.20 (a) Loading, (b) bending moment and (c) shear force diagram
                 of the substitute frame in the shorter direction (Figure 4.15) for
                 getting the maximum span (AB) 97

Figure 4.21 (a) Loading, (b) bending moment and (c) shear force diagram
of the substitute frame in the shorter direction (Figure 4.15) for
getting maximum span moment (BC) 98

Figure 4.22 (a) Loading, (b) bending moment (follow encircled value in
later stages of calculations) and (c) shear force diagram of the
substitute frame in the shorter direction (Figure 4.15) for dead
load only 99

Figure 4.23 Substitute frame at first floor level in the longer
direction (B–B)...101

Figure 4.24 Loading diagram of the substitute frame in the longer direction
(Figure 4.14) to get maximum support moment at 2.......................102

Figure 4.25 (a) Loading, (b) bending moment and (c) shear force diagram
of the substitute frame in the longer direction to get maximum
support moment at 2...107

Figure 4.26 (a) Loading, (b) bending moment and (c) shear force diagram
of the substitute frame in the longer direction to get maximum
support moment at 3...108

Figure 4.27 (a) Loading (follow encircled value in later stages of
calculations), (b) bending moment and (c) shear force diagram
of the substitute frame in the longer direction to get maximum
support moment at 4...109

Figure 4.28 (a) Loading, (b) bending moment and (c) shear force diagram
of the substitute frame in the longer direction to get maximum
support moment at 5...110

Figure 4.29 (a) Loading (point load in kN, distributed load in kN/m, length
in m), (b) bending moment and (c) shear force diagram of the
substitute frame in the longer direction to get maximum support
moment at 6..111

Figure 4.30 (a) Loading, (b) bending moment and (c) shear force diagram of
the substitute frame in the longer direction for getting maximum
span moment (1–2, 3–4, 5–6)..112

Figure 4.31 (a) Loading, (b) bending moment and (c) shear force diagram of
the substitute frame in the longer direction for getting maximum
span moment (2–3, 4–5, 6–7)..113

Figure 4.32 (a) Loading, (b) bending moment and (c) shear force
diagram of the substitute frame in the longer direction for dead
load only 114

Figure 4.33 Stress diagram for cantilever under horizontal loading....................116

Figure 4.34 Frame in the shorter direction (4–4) showing wind loads...............117

Figure 4.35 Frame in the longer direction (B–B) showing wind loads 128

Figure 4.36 Frame in the shorter direction showing seismic loads 131

Figure 4.37 Frame in the longer direction showing seismic loads 133

Figure 4.38 Position of column B4...151

Figure 4.39 Critical sections of one-way and two-way shear stresses.................168

Figure 4.40 Detail of top reinforcements of slab 169

Figure 4.41 Detail of bottom reinforcements of slab..170

Figure 4.42 Working details of reinforcements for beams of the
six-storied building..171

Figure 4.43 Schedule of reinforcements for beams of the six-storied
building..172

Figure 4.44 Working details of reinforcement for columns of the
six-storied building..173

Figure 4.45 Working details of reinforcement for foundations of the
six-storied building..174

Figure 4.46 Building plan..176

Figure 4.47 Plan showing wind angle and marking of external
wall faces..182

Figure 4.48 Reinforcement details of slab (top reinforcement only)
for 15-storied building..198

Figure 4.49 Reinforcement details of slab (bottom reinforcement only) for
15-storied building 199

Figure 4.50 Reinforcement details of beams for 15-storied building..................200

Figure 4.51 Reinforcement schedule of beams for 15-storied building 201

Figure 4.52 Reinforcement schedule of beams for 15-storied building 202

Figure 4.53 Reinforcement details of columns for 15-storied building...............203

Figure 4.54 Reinforcement details of pile caps for 15-storied building 204

Figure 4.55 Layout of foundation (pile caps) including grade beams and
piles and detail of reinforcements for 15-storied building 205

Figure 5.1 Plan of a six-storied building...208

Figure 5.2 Sectional elevation of a six-storied building 208

Figure 5.3 Plan of a six-storied building...214

Figure 5.4 Sectional elevation of a six-storied building 215

Figure 5.5 Wind load acting on a vertical wall (see figure 7.5 of
EN 1991.1.4: 2010) 219

Tables

Table 3.1	SPT value (N), density, percentage fine etc. of different layers of soil	36
Table 3.2	Values of N_{60} of different layers of soil	38
Table 3.3	Values of $(N_1)_{60}$ of different layers of soil	39
Table 3.4	Values of $CRR_{7.5}$ of different layers of soil	40
Table 3.5	Values of CSR, CRR and FoS	41
Table 3.6	Soil data showing shear wave values	42
Table 3.7	Values of CSR of different layers of soil	43
Table 3.8	Values of $CRR_{7.5}$ of different layers of soil	44
Table 3.9	Values of CRR and FoS of different layers of soil	44
Table 3.10	Soil testing data showing CPT values of different layers of soil	45
Table 3.11	Values of normalized dimensionless cone penetration resistance of different layers of soil	46
Table 3.12	Cyclic resistance ratio ($CRR_{7.5}$) of different layers of soil	47
Table 3.13	Cyclic resistance ratio (CRR) of different layers of soil	48
Table 3.14	Values of CSR and FoS of different layers of soil	49
Table 4.1	Slab panels are marked according to edge condition (as per table 26 of IS 456, 2000)	67
Table 4.2	Ultimate design bending moment for slab panels (as per table 22 of IS 456, 2000)	68
Table 4.3	Design calculations for reinforcement for slab panels	69
Table 4.4	Values of k_2 at different heights	71
Table 4.5	Values of k_1, k_2, k_3, V_b and V_z at different heights	72
Table 4.6	Values of V_z and p_z at different heights	72
Table 4.7	Design pressures at different heights	73
Table 4.8	Effective frontal area for each joint/node	76
Table 4.9	Wind load on nodes/joints (on the shorter face of the building)	77
Table 4.10	Wind load on nodes/joints (on the longer face of the building)	78
Table 4.11	External wind pressure coefficients	80
Table 4.12	Net wind pressure coefficients for different surfaces	80
Table 4.13	Wind pressure on wall when $\theta = 0°$	81
Table 4.14	Wind pressure on wall when $\theta = 90°$	81
Table 4.15	Weight of roof slab	84
Table 4.16	Weight of parapet walls, stair walls, water tank, etc. from roof level	85
Table 4.17	Weight of floor slabs	85
Table 4.18	Weight on account of walls, columns, etc. at a typical floor level	86
Table 4.19	Weight of walls, beams and column elements	86
Table 4.20	Distribution of lateral forces (inertial force) and shear forces at different floor levels	87

Table 4.21	Distribution of lateral forces (inertial force)	87
Table 4.22	Stiffness of members of the shorter frame (see Figure 4.15)	90
Table 4.23	Distribution factor (DF) at joints of the frame (see Figure 4.15)	90
Table 4.24	Fixed end moment for different type of load	92
Table 4.25	Fixed end moments (see Figure 4.19a and Table 4.24)	93
Table 4.26	Moment distribution method applied to short frame 4–4 (see Figure 4.4)	94
Table 4.27	Design bending moment (kNm) due to dead load only	100
Table 4.28	Design shear force (kN) due to dead load only	100
Table 4.29	Design bending moment (kNm) due to dead and live loads	100
Table 4.30	Design shear force (kN) due to dead and live loads	100
Table 4.31	Moment of inertia of beams and columns of the longer frame (B–B) (see Figure 4.23)	101
Table 4.32	Stiffness of the beams and columns of the longer frame (see Figure 4.23)	101
Table 4.33	Distribution factor at the joints of the longer frame (see Figure 4.23)	102
Table 4.34	Fixed end moments (see Figure 4.24)	104
Table 4.35	Moment distribution method applied for analysis of the long frame (B–B)	106
Table 4.36	Design bending moment (kNm) due to dead load only	115
Table 4.37	Design shear force (kN) due to dead load only	115
Table 4.38	Design bending moment (kNm) due to dead and live loads	115
Table 4.39	Design shear force (kN) due to dead and live loads	116
Table 4.40	Cantilever method of analysis	118
Table 4.41	Design column moment and shear force	127
Table 4.42	Design beam moment and shear force	127
Table 4.43	Axial force and bending moment of ground floor column under wind load in kN (refer to Table 4.40 at plinth level)	128
Table 4.44	Design column moment and shear force	129
Table 4.45	Design beam moment and shear force	130
Table 4.46	Axial force and bending moment of ground floor column under wind load in kN	130
Table 4.47	Design column moment and shear force	132
Table 4.48	Design beam moment and shear	133
Table 4.49	Axial force and bending moment of ground floor column under wind load in kN	133
Table 4.50	Design column moment and shear force	134
Table 4.51	Design beam moment and shear force	135
Table 4.52	Axial force and bending moment of ground floor column under wind load in kN	135
Table 4.53	Bending moment due to dead and live loads (kNm) (same as Table 4.29)	135
Table 4.54	Shear force due to dead and live loads (kN) (same as Table 4.30)	136

Table 4.55 Bending moment due to dead load only (kNm) (same as
Table 4.27)... 136

Table 4.56 Shear force due to dead load only (kN) (same as Table 4.28).......... 136

Table 4.57 Bending moment due to dead and live loads (kNm) (same as
Table 4.38)... 136

Table 4.58 Shear force due to dead and live loads (kN) (same as
Table 4.39)... 137

Table 4.59 Bending moment due to dead load only (kNm) (same as
Table 4.36)... 137

Table 4.60 Shear force due to dead load only (kN) (same as Table 4.37).......... 137

Table 4.61 Bending moment and shear force of first floor beams due to
wind load (kN) (same as Table 4.42)... 138

Table 4.62 Axial force of ground floor column due to wind load (kN)
(same as Table 4.43)... 138

Table 4.63 Bending moment and shear force of first floor beams due to
wind load (kN) (same as Table 4.45)... 138

Table 4.64 Axial force of ground floor column for wind load (kN)
(same as Table 4.46)... 139

Table 4.65 Bending moment and shear force of first floor beams for
seismic load (kNm) (same as Table 4.48) 139

Table 4.66 Axial force of ground floor column for seismic load (kN)
(same as Table 4.49)... 140

Table 4.67 Bending moment and shear force of first floor beams due to
seismic load (kN) (same as Table 4.51)... 140

Table 4.68 Axial force of ground floor column for seismic load (kN)
(same as Table 4.52)... 140

Table 4.69 Load combination.. 141

Table 4.70 Load combination of bending moments for the shorter frame
beam (4–4) (first floor) 141

Table 4.71 Load combination of shear forces for the shorter frame beam
(4–4) (first floor).. 143

Table 4.72 Load combination of bending moments for the longer frame
beam (first floor).. 144

Table 4.73 Load combination of shear forces for the longer frame beam 145

Table 4.74a Design of shorter frame beams (4–4) for longitudinal
reinforcements (see Tables 4.70 and 4.71) 147

Table 4.74b Design of shear reinforcements of shorter frame beams (4–4) 148

Table 4.74c Design of longer frame beams (B–B) for longitudinal
reinforcements (see Tables 4.72 and 4.73) 149

Table 4.74d Design of shear reinforcements of longer frame
beams (B–B)... 150

Table 4.75 Load combination for design of columns at ground
floor level.. 154

Table 4.76a Calculation of final design values of P_u, M_{ux} and M_{uy} for column
B_4 (first trial)... 155

Table 4.76b Design calculations for column B$_4$ (first trial) 156
Table 4.76c Design calculations for column B$_4$ (first trial) 158
Table 4.77a Calculation of final design values of P$_u$, M$_{ux}$ and
                       M$_{uy}$ for column B$_4$ 160
Table 4.77b Final design calculations for column B4 161
Table 4.77c Final design calculations for column B4 162
Table 4.78   Final design calculations of combined check for column B4 164
Table 4.79   Unfactored axial force (P) and bending moment (M$_x$ and M$_y$) or
                       different load combinations 165
Table 4.80   Terrain factor (k$_2$) at different heights 178
Table 4.81   Design wind speed at different heights 178
Table 4.82   Wind pressure at different heights 179
Table 4.83   Design wind pressure 179
Table 4.84   Wind force along the X direction 181
Table 4.85   Wind force along the Z direction 181
Table 4.86   External pressure coefficients 182
Table 4.87   Net pressure coefficients considering external and internal
                       pressure coefficients 183
Table 4.88   Wind force calculation 183
Table 4.89   Natural frequencies 184
Table 4.90   Calculation of p$_d$ 187
Table 4.91   Calculation of turbulence intensity and roughness factor 188
Table 4.92   Effective turbulence length, background factor, height
                       factor, etc. 189
Table 4.93   Peak factor, factor to account for the second-order
                       turbulence intensity 190
Table 4.94   Size reduction factor 190
Table 4.95   Spectrum turbulence and effective reduced frequency 191
Table 4.96   Gust factor (G) 192
Table 4.97   Wind forces at different levels along heights 193
Table 4.98   Cross-wind force spectrum 194
Table 4.99   Values of M$_c$ and F$_{zc}$ (forces considering per unit height
                       and per unit width) 196
Table 5.1   Basic data considered for wind load analysis 209
Table 5.2   Velocity pressure coefficient (K$_z$) along heights 211
Table 5.3   Velocity pressure (q) along heights 211
Table 5.4   Calculation of external pressure coefficient (C$_p$) for wall 212
Table 5.5   Design wind pressure on the windward wall 212
Table 5.6   Wind pressure on the leeward wall 212
Table 5.7   Design wind pressure on the side walls 213
Table 5.8   Basic data for wind load assessment 215
Table 5.9   External wind pressure coefficients 219
Table 5.10   Wind pressure on wall 220
Table 5.11   Comparison of basic parameters of wind and seismic load
                       assessment 222

Preface

Reinforced concrete as a construction material provides enormous architectural freedom. The basic concepts for the design of Reinforced Concrete structural elements are, fundamentally, more or less the same, but students and practicing engineers have access to a number of computer software packages, hundreds of textbooks, articles and research papers and large amounts of online information. It is difficult to recommend the best and most appropriate design resources. The present engineering education system focuses more on computer-based mathematical models without understanding their shortcomings. However, such analyses would be more powerful and useful if they had been developed in the light of realistic engineering concepts. Intuitive skill and experience, along with computer-aided mathematical analysis, need to be given due attention. The design concept needs to be cost-effective and better than its alternative if it is to be accepted.

This book explains the wind and seismic design issues of Reinforced Concrete buildings in brief and provides design examples based on the recommendations of the latest Indian Standard (IS) codes, which are essential and mandatory documents for industrial design and in order to achieve an acceptable common platform of understanding. It also provides detailed working drawings; several such typical design examples of buildings are given. The book provides the basic insights necessary for the effective development of a design. The most intricate issues of Reinforced Concrete design are discussed, supplemented by a number of real-life examples for deeper understanding of the subject.

Guidelines are presented for evaluating the acceptability of wind-induced motions of tall buildings. A design methodology for members or the structure as a whole to deform well beyond their elastic limits, which is essential under seismic excitation, is discussed in depth. Detailed considerations for such nonelastic behavior, in order to formulate simple procedures to accommodate design objectives that receive due attention in the code provisions, are also critically discussed.

The target readers of this book are the practicing structural engineers and architects, students and teachers of Civil engineering and Architecture, striving to understand the design of wind- and earthquake-resistant Reinforced Concrete buildings. This book explains the concepts on behavior of Reinforced Concrete buildings against wind and earthquake forces. It is an attempt to respond to some of the frequently asked questions by architects and structural engineers regarding the design of reinforced concrete buildings against wind and earthquake forces. Detail design calculations and reinforcement detailing as per recommendation of different relevant codes of practices have been furnished. Reinforcement detailing in the form of working drawings are also included so as to reduce the gap of understanding between different groups of professionals in structural engineering. The book is intended to serve as a comprehensive reference for the wind and earthquake-resisting design calculations and details of Reinforced Concrete buildings inline with the latest IS codes. It is a user-friendly complete package for all who deal with Reinforced Concrete design, explaining the wind and seismic design issues of Reinforced Concrete buildings and providing numerical design examples, along with working drawings based on the recommendations of the latest IS codes.

Professor (Dr.) Somnath Ghosh
Dr. Arundeb Gupta

Acknowledgements

We would like to take this opportunity to remember the sacrifice made by and encouragement received from our wives, Shaswati and Manasi, as well as our children: Kushal, Madhurima and Swarnabha.

We are thankful to a number of persons who helped us in the preparation of the manuscript.

We are extremely thankful to Dr. Gagandeep Singh and Mr. Lakshay Gaba of CRC Press, Taylor & Francis, for their guidance and help in writing this book.

Professor (Dr.) Somnath Ghosh
Dr. Arundeb Gupta

Authors

Somnath Ghosh is serving as Professor in the department of Civil Engineering Department at Jadavpur University in Kolkata, India. He was Dean of the Engineering Faculty and Head of the Civil Engineering Department at Jadavpur University, India. Dr. Ghosh obtained his B.E. in Civil Engineering from Jadavpur University and his M.Tech and Ph.D from the Indian Institute of Technology Kharagpur, India. He is a member and chartered engineer of the Institute of Structural Engineers in the United Kingdom. He has carried out research in the United States and Australia and delivered invited lectures in the United Kingdom, Australia, Singapore, Malaysia, Thailand and the United States. He has also delivered a huge number of lectures as a resource person in different Indian Institutes of Technology, National Institutes of Technology and universities. He has served as an expert member on several occasions for many institutes and universities. He has served as a member of several high-powered committees in the All India Council for Technical Education, the University Grants Commission, the Council of Scientific and Industrial Research, the Union Public Service Commission, etc. and at Jadavpur University level. He has contributed significantly in the areas of structural engineering and materials. He has also been a structural consultant to a number of key projects at the national level. Based on his research works, Dr. Ghosh has published a number of papers in peer-reviewed national and international journals and also published six monographs at international level. Apart from his research activities, Dr. Ghosh has demonstrated his technical skill by providing advice on industrial projects, and these have been implemented successfully. The repair and restoration techniques adopted for the earthquake-damaged structures of Kandla Special Economic Zone through his expertise deserve special mention. His other noteworthy contribution is the restoration of the earthquake-damaged assembly building in Sikkim. In addition, his skill in computer-aided structural analysis has been demonstrated through the design of a 52-meter tall Buddha statue at the top of a hill at Rabangla, Namchi, Sikkim, and a cricket stadium at Guwahati, Assam. His selection as country head for a division of a multinational company in Nigeria, speaks volumes about his administrative abilities as well as his academic skill and expertise;

Arundeb Gupta serves as Principal Structural Consultant for Skematic Consultants, in Kolkata, India. He obtained his B.E. and Ph.D in Civil engineering from Jadavpur University, India and has now built up around 30 years' industrial experience. He has designed several critical structures in highly seismic hilly terrain, as well as special structures, such as multipurpose cyclone shelters in coastal areas. Dr. Gupta has also carried out the restoration and designed the retrofitting of several earthquake-damaged structures. He has completed a large number of projects in health care, education, heavy industry, residential sectors, etc. at the national level. Dr. Gupta has published several papers in peer-reviewed national and international journals and he also serves as guest faculty in the Civil Engineering Department of Jadavpur University, India.

Notation

P_d	design wind pressure
K_d	wind directionality factor
K_a	area averaging factor
K_c	combination factor
p_z	design wind pressure
V_z	design wind speed
k_1	risk coefficient
k_2	height and terrain factor
k_3	topography factor
k_4	importance factor cyclonic region
V_b	basic wind speed
C_{pe}	external pressure coefficient
C_{pi}	internal pressure coefficient
A	surface area of structural element or cladding unit
C_f	force coefficient
A_e	effective frontal area
H	height of the building
B	width of the building
L	length of the building
F_z	design peak along-wind load on the building structure at any height z
$C_{f,z}$	drag force coefficient of the building structure corresponding to the area A_z
p_d	design hourly mean wind pressure
V_{zd}	design hourly mean wind speed at height z, in m/s
A_z	effective frontal area of the building structure at any height z, in m²
G	gust factor
r	roughness factor
g_v	peak factor for upwind velocity fluctuation
B_s	background factor
b_{sh}	average breadth of the building/structure
L_h	measure of effective turbulence length scale at the height, h, in m
$Ø$	factor to account for the second order turbulence intensity
$I_{h,i}$	turbulence intensity at height h in terrain category i
H_s	height factor for resonance response
S	size reduction factor
E	spectrum of turbulence in the approaching wind stream
N	effective reduced frequency
f_a	first mode natural frequency of the building/structure in along-wind direction, in Hz
V_{hd}	design hourly mean wind speed at height, h in m/s
$ß$	damping coefficient of the building/structure, for Reinforced Concrete structure
g_r	peak factor for resonant response

k_{21}	hourly mean wind speed factor for terrain category 1
z	height or distance above the ground
z_{0i}	aerodynamic roughness height for ith terrain
$F_{z,c}$	across-wind load per unit height at height z
K	mode shape power exponent for representation of the fundamental mode shape
f_c	first mode natural frequency of the building/structure in across-wind direction in Hz
b	breadth of the structure normal to the wind in m
p_h	hourly mean wind pressure at height h, in Pa
C_{fs}	cross wind force spectrum coefficient
M_b	P-wave magnitude
M_s	surface-wave magnitude
M_0	seismic moment
M_w	moment magnitude
M_L	Richter magnitude
D	average fault displacement
A	total area of the fault surface
M	average rigidity with respect to the shearing forces of the rocks
A_h	design horizontal seismic coefficient
Z	zone factor
I	importance factor
R	response reduction factor
S_a/g	design acceleration coefficient
T_a	fundamental translational natural period
V_B	base shear
W	seismic weight
Q_i	design lateral force at ith floor level
w_i	seismic weight of ith floor level
h_i	height of ith floor measured from base
n	number of floors including roof
M_k	modal mass of kth mode
P_k	modal participation factors of kth mode
Q_k	design lateral load for ith floor mode k
V_{ik}	story shear for ith floor mode k
p_t	percentage of tension steel
d	effective depth
D	overall depth
l_x	maximum shorter span of a slab panel
l_y	longer span of a slab panel
α_x	positive shorter span coefficient of the bending moment of a slab panel
α_x'	negative shorter span coefficient of the bending moment of a slab panel
α_y	positive longer span coefficient of the bending moment of a slab panel
α_y'	negative longer span coefficient of the bending moment of a slab panel
M_u	ultimate design bending moment
M_{ul}	ultimate uniaxial bending moment capacity in the presence of axial compressive load

P	percentage of steel
V_u	ultimate design shear force
A_{st}	area of tension steel
S	spacing of reinforcement in slab
WL–X	wind force acting along longer direction on shorter face
WL–Y	wind force acting along shorter direction on longer face
EQ–X	seismic force acting along longer direction on shorter face
EQ–Y	wind force acting along longer direction on shorter face
f_y	0.2 percent proof stress of HYSD (high yielding strength deformed) bars
f_{ck}	characteristic strength of concrete
A_{sv}	total cross-sectional area of stirrup legs effective in shear
S_v	stirrup spacing along the length of the member
d′	effective cover
e	eccentricity

1 Introduction

1.1 PREAMBLE

Huge advancement has taken place in structural engineering but still there is no easy and direct answer to the question: "What is the requirement of a building structure to resist wind and earthquake forces so as to get reasonable serviceability at an optimum cost?" Few thumb rules and a huge amount of computer-aided rigorous analyses, are available. Available rigorous analyses are backed by a number of software also. It is possible to analyze the building structures to appreciate the actual behavior of the structure to a great extent. At the preliminary stage, type of structural systems, approximate sizes of structural members etc. are considered by a structural engineer based on thumb rules and experience. A seismic resistant structure necessitates integration of different structural elements which will resist inertia forces and transfer them along force paths prefixed by the structural engineer to the foundation. The suspended floors provide adequate diaphragm strength and stiffness. But the structural adequacy of a diaphragm warrants special attention if the distance between vertical/horizontal members (beams and columns) are higher. Presence of cutouts weaken it and reduce its spanning ability where planning of vertical and horizontal members needs a carefully attention. Ductility and Hysteresis are two important and interrelated aspects and a very clear understanding is required for a structural engineer. Redundancy and collapse pattern play an important role in developing design process. Behavior of Reinforced Concrete structure is not very straightforward. However, negative influences may often be avoided by creative structural configuration. In the next section, a brief discussion will be made to appreciate the subject as a whole.

1.2 A FEW IMPORTANT ASPECTS OF STRUCTURAL DESIGN

Reinforced Concrete structures is not perfectly elastic even at lower stress level. At higher stress level, it undergoes cracking etc. Stiffness of Reinforced Concrete elements will decrease appreciably and deformations will increase drastically but it will not collapse immediately if the Reinforced Concrete structures made sufficiently ductile. It is bit difficult to understand behavior of Reinforced Concrete structure under dynamic loads like subjected to wind, seismic forces etc. However, buildings should be designed incorporating the ductile detailing to sustain these loads undergoing larger deformations but no collapse. Total energy level of ductile Reinforced Concrete may be very high and ultimately design will be reasonable and economic ensuring less damage as well as safety of the structure. It is essential to understand both strength as well as serviceability aspects. Ductility and Hysteresis are two important and interrelated aspects and a very clear understanding is required for a structural Engineer. Redundancy of structure needs to be understood to appreciate sequence of failure and to assess overall ductility of the structure. In the next section, these aspects are discussed to appreciate its design application made in Chapter 3 through design examples and detailing.

If the center of mass and center of stiffness of a building do not coincide on plan at any floor level, the building undergoes torsion about its vertical axis. Generally, for a symmetrical building, the stiffness and mass are found symmetrical on plan. In the inelastic stage if a weaker members reach their limiting strength, stiffness of those members get deteriorated compare to other stronger members which have not deteriorated to that extent, results in a stiffness asymmetry problem. At this stage, center of mass and center of stiffness do not coincide. This is a common problem at inelastic stage. This area needs more investigation.

1.2.1 STRENGTH AND SERVICEABILITY

It is essential to understand inelastic behavior of Reinforced Concrete structures to ensure safety and good performance under earthquake shaking. The hysteresis model is extremely important to appreciate the strength and stiffness degradation, ductility, and energy dissipation capacity of structures. Serviceability includes factors such as durability, overall stability, fire resistance, deflection, cracking and excessive vibration. A structurally perfect tall building may sway to a great extent which causes the sickness to the occupants. This building will not collapse even it exceeded its serviceability limit state. Excessive deflection, vibration and local deformation need to be avoided. Building codes provide serviceability limits, ensuring public and occupant safety. Generally, the economic cost of a structure is the priority in a project and structure design becomes the point of attraction and attention. Many difficult task becomes simpler for the availability of design software. Structural optimization may be made, without exceeding the design criteria, i.e., strength, serviceability and stability, to achieve a better economic solution. Nevertheless this area is too broad to be discussed properly, needs detail discussion.

1.2.2 DUCTILITY AND HYSTERESIS

It may said that ductility of a building is its capacity to accommodate large lateral deformations, quantified primarily with a ratio of maximum deformation sustained just prior to collapse or significant loss of strength to the deformation at yield. Large inelastic deformation capacity without significant loss of strength capacity, is the most important observation. Buildings are designed and detailed to develop favorable failure mechanisms which involves specified lateral strength and reasonable stiffness. Ductility helps in dissipating input earthquake energy through hysteretic behavior. Ductile chain design concept in building as per capacity design, is extremely popular.

When a brittle chain alone is pulled on either side, it would break suddenly. Among the chain, the weakest link would break first. If we make the weakest link as the ductile one, it provides more elongation though failure load may be more or less same, i.e., gains ductile. Similar concept is applied in case building design. Inertia forces due to seismic activity are transmitted from the floors to the beams then to the columns. The stability of the building is affected mostly failure of the column rather than the failure of beams. Therefore, it is better to make beams as weak ductile links than columns. This concept known as "**Strong column-Weak beam concept**". The structure gains a larger ductility other than the contributions of material ductility is due to special joint detailing or making stirrups closer etc. as recommended in different design codes of practices. The structure are going through plastic stage where stiffness decreases appreciably and deformation is drastically increasing. It should be ensured that the structure sustains these loads undergoing smaller damage (repairable) but no collapse. Ductility plays an important role to achieve this target. A ductile building can withstand inelastic actions without losing stability, avoiding collapse and undue loss of strength at deformation levels beyond the elastic limit. Ductility of a building is assessed through different concepts of ductility, i.e., global ductility, member ductility, sectional ductility and material ductility. These ductility concepts are interrelated. The stirrups are primarily designed for resisting shear force but provides confinement to core concrete also. Similarly, lateral ties are provided in columns to avoid buckling of longitudinal reinforcement as well as carries shear force, but provides confinement to core concrete also. The core concrete is prevented from dilating in the transverse direction thereby enhancing its peak strength and ultimate strain capacities. The use of closer spacing of stirrups in beams or closer spacing of lateral ties, makes confining pressure more uniform and effective. The concrete, a brittle material gains ductility when provided with closer stirrups/lateral ties to confine core concrete. Steel by nature is far more ductile than even confined concrete. But, it must be ensured that such steels have at least the prescribed minimum elongation specified in different seismic design codes; for instance, as per IS 13920, steels used in earthquake-resistant constructions should have at least 14.5% elongation at fracture. Material ductility is reflected through moment-curvature relationship. Sectional ductility is assessed from moment-curvature relationship. If a building behaves elastically during earthquake shaking, no damage is occurred. If there is a requirement imposed on the structure to undergo inelastic action, i.e., demand for the ductility, then it will undergo some inelastic deformation beyond yield deformation but no collapse, allowing small pre-decided damages. The global ductility of the structure can be obtained from the ratio of maximum displacement of the building to its yield displacement.

Different types of failure occurs in Reinforced Concrete structures, i.e., shear failure, bond slip failure, flexural over-reinforced failure, flexural under-reinforced failure, torsional failure etc. In general, flexural under-reinforced failure is preferred because where the Reinforced Concrete member stretches in flexure on the tension side without any failure of compression concrete and exploits the ductility of the steel bars. This plastic action spreads over a small length of the member forming plastic hinge. The effect of enhanced ultimate strain of concrete is reflected in the moment-curvature relation of the section. The curvature ductility is significantly increased. However, the moment carrying capacity of the section is governed by tensile capacity of the steel and not the strength of the concrete. The maximum strain capacity of concrete is increased for confined concrete and as a result, curvature ductility is also enhanced and ultimately reflected in the global load-deformation response of structures. Pushover responses of the building with and without transverse confining reinforcement, also changes appreciably. The global drift capacity of the building is significantly increased with additional confining reinforcement. Reinforced Concrete columns must not be designed to carry axial compression load above the balanced condition as this leads to brittle compression failure. Curvature ductility is significantly enhanced due to imposed confinement below the balanced level and will prevent brittle collapse of columns in case of extreme shaking due to earthquake. Quantification of ductility is still not common. Design for ductility is normally done through prescriptive advices, like recommending certain specific reinforcement detailing, on structural configurations, imposing material specifications, demanding an acceptable sequencing of possible failure modes. Finally, the level of ductility of the building structure or a structural element is obtained by applying the **"Pushover analysis"**.

It is a common understanding to achieve economy and safety for building structures under earthquake shaking, let the structure undergo small damages either due to plasticity, fracture, cracking etc., ensuring its strength to carry the vertical load to prevent collapse. Lateral ties are provided to confine concrete and to avoid buckling of longitudinal reinforcement. It would enable the column to continue taking vertical loads even if it is subjected to cracking/ concrete crushing/yielding of steel reinforcement, but the stiffness of the structural elements/structure would fall drastically. It may be noted here that, if the structure had remained elastic there would be higher internal forces and total base shear, i.e., the sum of internal shear forces in all the columns. Therefore, incorporation of ductility in the material, allowed the structure to get damaged, thus reduces internal forces. Provision in this regard is provided in building codes worldwide. The ductility factor depends on upon the lateral structural system provided. Another aspect is, determining the failure mechanism of members. The idea is to force a member/structure to undergo failure in a ductile manner.

1.2.3   REDUNDANCY

Redundancy of a structural system, is a non-independent concept and has an indirect effect on ductility. It is a common understanding that any increase in structural redundancy provides a desirable property to deal with more effectively against earthquake shaking and can reduce structural sensitivity against undesirable loads. Hence, a discussion is necessary clarify the role of redundancy under earthquake shaking and to distinguish the role of redundancy from total capacity. Increase in redundancy would not always lead to significant improvement in structural behavior under seismic excitation. It is better to examine the effects of redundancy on the ductility.

1.3   ARCHITECTURAL REQUIREMENTS

Architectural concept and Structural design, are both equally important. A good integration of structural considerations with the realization of the architectural concept, help to decide architectural form and planning. Cost-effective and planning friendly structural solutions, need to achieved, which will provide best possible aesthetic value of the building. Architectural design and Structural design should proceed together from the beginning of the project without compromising architectural ideas as well as safety of the building at a reasonable cost. Overall geometry, Structural systems, and Load paths are three basic important aspects. Load paths, through different structural elements, till reaches finally reaches to soil through the foundation. Inertia forces developed due to earthquake shaking are significantly influenced by overall geometry of the building, i.e., shape and aspect ratio of the building plan and slenderness ratio of the building. Convex geometry of the buildings have direct load paths for transferring forces to its base, while concave geometry of the buildings takes indirect load paths that result in stress concentrations at points where load paths are curved one. Buildings with convex and simple plan are referred for better seismic performance.

Configuration of the building alone is the major contributor to building failure. Generally, Architects decides the overall form or massing of a building. It is important for the architects having a sound appreciation of good and poor structural stability as well as seismic configuration. It is essential to appreciate horizontal configuration as well as vertical configuration together. Configuration issues, perhaps the most important issue to be resolved at the beginning to avoid conflict between architects and structural engineers. Architects are in favor of more stimulating plan forms and the clients are attracted toward non-rectangular plan forms. However, conflicts need to be resolved to ensure safety of the structure, paying respect to the ideas of architect and client.

So far as dynamic behavior of the structure, it is desirable to have pure translation modes as the lower modes of vibration and push other modes to the higher level. Correct assessment of time period is extremely important. Undesirable diagonal translation and torsional modes are occurring due to lack of symmetry of the plan shape. For buildings having complex shapes, particularly with projections or re-entrant corners, undergo some different modes of vibration in addition to translation modes, diagonal or torsional modes etc. It is generally described with wagging of a dog's tail, i.e., only long projection vibrates, keeping the remaining part of the building more or less still. It is like the dog's body remains still when its tail wags. It induce high stress concentration at the re-entrant corners that may cause significant damage of the building structures due to earthquake shaking (refer to Figure 1.1). This effect is more pronounced in buildings with T and L shaped plan of the building. But, in both buildings, torsional mode is predominant. It is better to make two separate rectangular buildings with a construction joint to avoid torsional effect to a great extent. Translational modes of vibration are predominant in regular buildings or buildings with small projections. Similarly, in buildings with V, Y or X shaped plan, undesirable modes of vibrations can be avoided in buildings having small projections. It is not good to have buildings with large plan aspect ratio and buildings with large projections. If the projected length is more than a certain percentage of overall length of building, then it is considered as a serious irregularities. In planning this type of situation should be avoided as chances of damage due to earthquake shaking is much higher than a regular building. If this percentage is more than 15%, IS 1893, provided separate recommendation to deal with. The structure may be subdivided into

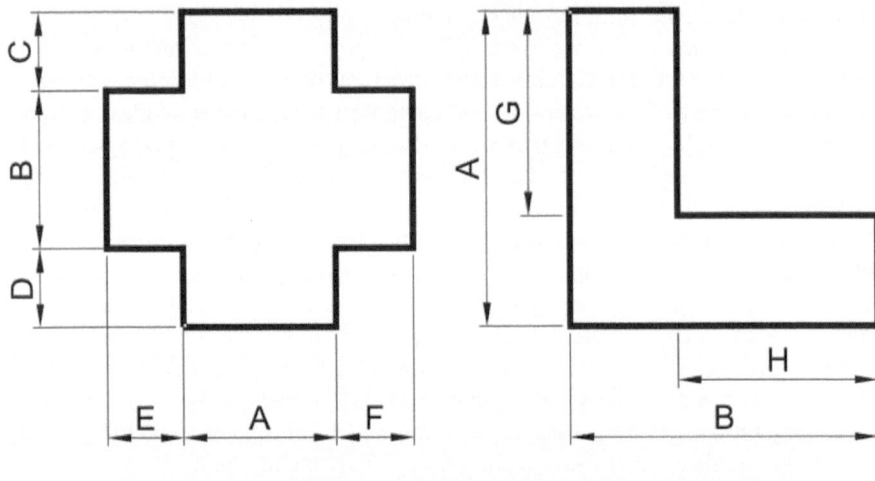

PLAN OF RE-ENTRANT CORNERS

FIGURE 1.1   Re-entrant corner

dynamically stable independent units. Irregularity may be accepted to some extent in order to have the aesthetics according to architects dream provided more time consuming rigorous, design methodology is adopted to ensure safety and accepting increase in total cost. It is observed that the performance of buildings with irregular configuration is extremely good compare to a regular structures. Codes are generally suggesting stronger structural connections and members compare to normal one and lead to more cost. Frankly speaking, in general, codes should prohibits extremely irregular structures in high seismic zone.

Geometrical plan complexity may be enhanced due to introduction of large floor openings/cut-outs (refer to Figure 1.2). For introducing daylight or accentuating spatial variety or hierarchy, it may be one of the clients requirements for different factors including, i.e., architectural design concept, the desirability of introducing natural light and ventilation etc. Reinforced Concrete slab may be considered as rigid diaphragm. If the slab is having excessive cutout, problem arises several folds. If the slab panels having less cutout/ openings, the behavior of the slab panel depends upon the position and size of the opening/cutout. This is also a type of the irregularities. Different codes have their a particular recommendations to handle structural behavior.

Few horizontal irregularities may be (i) Torsional and extreme torsional (ii) Reentrant corner (iii) Diaphragm discontinuity (iv) Out-of-plan offsets and (v) Nonparallel systems (refer Figures 1.3 and 1.4).

There are different kinds of irregularities as described in IS 1893 like plan irregularities (includes torsional irregularities etc.), vertical irregularities (includes stiffness irregularities, mass irregularities etc.). As described in IS 1893, that if the maximum horizontal displacement of one end to minimum horizontal displacement of other end ratio is more than 1.5 then the building is irregular from torsion point of view. If

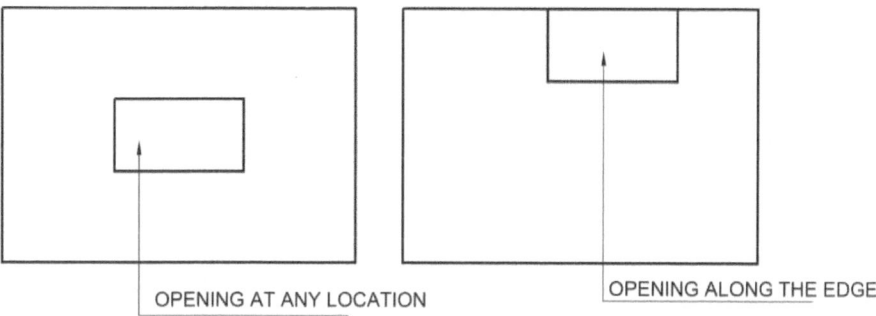

OPENING AT ANY LOCATION

OPENING ALONG THE EDGE

SLAB PANELS WITH CUT-OUT

FIGURE 1.2 Slab panels with cutouts

ELEVATION
CUT OF PLANE OFFSETS IN VERTICAL ELEMENTS

FIGURE 1.3 Out-of-plan offsets

the ratio lies between 1.5 and 2 then it is to be ascertained by changing the building configuration that fundamental torsional mode of oscillation of structure shall be less than the first two translational mode in each principal plan direction. If the ratio is more than 2.0 then building configuration to be revised. The structure should be designed with the load combination recommended by code.

Twist about the center of stiffness takes place, if the center of mass of a building is not coinciding with the center of stiffness and this affects furthermost column from the center of stiffness most severely. Column member will undergo large horizontal

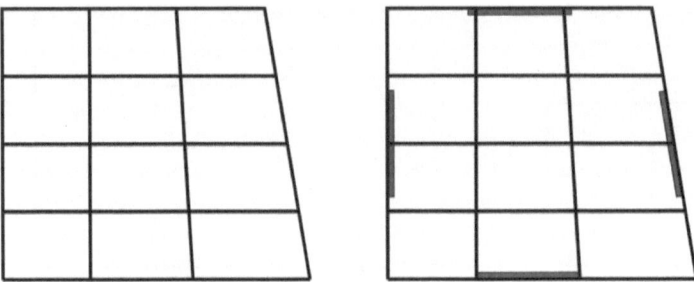

MOMENT RESISTING FRAME AND
MOMENT RESISTING FRAME WITH SHEAR WALL

FIGURE 1.4   A non-parallel system

deflections, sometimes damaging to a great extent lead to collapse under the influence of vertical gravity load. Torsion effect is found as common, observed after post-earthquake survey of Reinforced Concrete buildings. Therefore, the building should be planned in such a way, so that the distance between center of stiffness and center of mass, is minimum. Torsional motions within the ground movement is also possible. Different codes of practices recommended a minimum design eccentricity to account for the said unavoidable effects. The horizontal off-set between each line of structural elements should be as large as possible to maximize both the torsion strength and stiffness. Shear walls provides no considerable resistance under this torsional effect rather warp, bends about their weak axes. The rotation of the floor diaphragm also occurs. A Reinforced Concrete diaphragm is usually very stiff and strong in its plane and considers as a rigid unit. Distance between pairs of walls provide higher resistance against torsion. Replace the shear walls with one- or multi-bay moment frames and the principles outlined above still apply. Buildings are normally suffer from seismic damage due to re-entrant corners. Re-entrant geometries can take many shapes and their potential for damage resulting from the different dynamic performance of each wing.

Inertia forces are induced, normally at the floor levels where the mass is large, during earthquake shaking and distributed to columns/beams/structural walls. Inertia forces are distributed to columns/beams/structural walls in proportion to their stiffness/load resisting capacities, provided floor slabs do not deform to a great extent in their own plane, lead to extra loading on the members, causing damage. It is observed that the maximum displacement of the floor/diaphragm increases with increase in aspect ratio. IS 1893 recommended to restrict maximum lateral in plane displacement of diaphragm/floor at any point to within 1.5 times the average displacements of the entire diaphragm for good performance of buildings in buildings with plan aspect ratio not greater than 4, against earthquake shaking. It is required to distribute the inertia forces through all the columns/beams/structural walls to establish a direct load paths. Due to inelastic actions, columns/beams/structural walls lose stiffness and stiffness distribution is changed and stiffness eccentricity is increased, which affects the performance of the building during earthquake shaking. The problem worsen, for building with large plan, large openings/cut-outs, resulting in uneven distribution of inertia forces to columns/structural walls, during earthquake shaking. This type of irregularity should be avoided/minimized. The maximum displacement at the middle of the diaphragm increases with increase in area of opening. IS 1893 recommended to restrict maximum opening to 50% of diaphragm area. This is expected to limit the in-plane flexibility of the diaphragm and ensure in-plane lateral displacement of diaphragm at any point to within 1.5 times the average displacements of the entire diaphragm. Large lateral displacements cause significant non-structural damage, structural damage and even second order P-Δ effects that lead to collapse of buildings. IS 1893 recommend that inter-storey drift under design earthquake forces be restricted to 0.4 percent of storey height. Thus, maximum damage is expected to be confined to the first few storeys in buildings.

If a building is designed with conventional construction materials and compared with a similar building except the number of stories are more, maintaining the same total weight by using lightweight construction materials, there will be a reduction of acceleration response of the building. As the building height increases in turn time period increases and consequently total design inertia force decreases. However, the reduction in inertia force is insufficient to counteract the increased bending moments due to the greater height of the building.

1.4 LATERAL LOAD-RESISTING SYSTEM

1.4.1 Subsystems and Components

The subsystems or components of the tall building structural systems are essentially the following.

- floor systems
- vertical load-resisting systems
- lateral load-resisting systems
- connections
- energy dissipation systems and damping

These are broadly defined as follows:

- moment-resisting frames
- shear wall frame systems
- shear truss/outrigger braced systems
- framed tubes
- tube-in-tube systems with interior columns
- bundled tubes
- truss tubes without interior columns
- modular tubes

The structural system should be able to carry different types of loads, such as gravity, lateral, temperature, blast and impact loads. The drift of the tower should be kept within limits, such as H/500.

1.4.2 Moment-Resisting Frames, Braced Frames and Shear Walls

The most popular vertical structural systems to resist horizontal seismic forces are shear walls, moment-resisting frames and braced frames. These systems are capable of resisting gravitational forces and providing horizontal resistance. These are, basically, acting as a vertical cantilever more or less fixed at foundation level, and primarily resist horizontal forces due to earthquakes, wind, etc.

Building structures are generally a three dimensional framework of structural elements which acts integrally to resist loads. Vertical elements, diaphragms etc. are commonly the resisting systems adopted. Moment-resisting frames are generally selected as one of the seismic resisting system to allow flexibility of architectural planning. Moment-resisting frames are commonly used as lateral resisting systems when sufficient ductility and deformability demand need to be fulfilled. Specially moment-resisting frame (SMRF) that must be specifically detailed to provide ductile behavior and comply with the provisions of the guideline of the codes. Special Moment-Resisting Frame (SMRF) is a moment-resisting frame that fulfills special detailing IS 13920 and IS 4326. Ordinary moment-resisting frame (OMRF) is a moment-resisting frame that does not fulfill the special detailing requirements for ductile behavior of the frame. A highly ductile building frame of high degree of redundancy, which can allow freedom in architectural planning of internal spaces etc. Its flexibility and associated long period may serve to detune the structure from the forcing motions on stiff soil or rock sites. However, poorly designed, moment-resisting frames have been observed to fail catastrophically during earthquakes, due to formation of weak stories, failure of beam–column joints etc. Beam–column joints are zones of high stress concentration and therefore needs special attention and care during designing it.

Proportioning and detailing requirements of a Specially Moment Frames need to done carefully so that it becomes capable of multiple cycles of inelastic stresses without major loss of strength. Moment-resisting frames may be considered as the structures with a satisfactory behavior under severe earthquakes. These frames can provide a large number of dissipative zones, where plastic hinges form with potentially high dissipation capacity. In order to maximize the energy dissipation capacity, it ensures failure as a whole. Moment-resisting frames can provide different level of strength and ductility, if recommendations of code and expert opinions are strictly followed.

Reinforced concrete Specially Moment Frames are made, to improve performance of the beam, columns and joint, to achieve more overall ductility of the structure. It is required to achieve strong column- weak beam structural system which ensures ductile response in the yielding regions. Non-structural elements, in fills need to be detailed in a way, so that the target behavior could be achieved. Failure of a column is of greater impact than failure of a beam. Building codes recommended that columns shall be stronger than the beams. This strong column –weak beam concept is the fundamental approach to achieve safety during strong earthquake ground shaking. Many codes adopts the strong column-weak beam concept and provided guidelines to achieved by keeping the sum of column moment strengths exceed the sum of beam moment strengths at each joints. Non-ductile failures can be avoided using a capacity design approach. Shear failure, especially in columns, is relatively brittle and need to be avoided. A capacity design approach that requires the design shear strength should be at least equal to the shear that occurs when yielded sections reach near to ultimate moment capacity. The strength of lap/spliced longitudinal reinforcement loses cover during earthquake shaking. Therefore, lap/splices should be located away from sections of maximum bending moment, i.e., away from beam-column junction and closed ties/stirrups need to be provided to confine the lap/splice zone. Different codes restricts diameter of reinforcement also. Bars anchored in exterior joints must have hooks extended as recommended by different codes of practices. Different requirements apply to interior and exterior joints. The cracked stiffness of the beams, columns, and joints must be appropriately considered while analyzing Special Moment Frame which provides reasonable building periods and thus base shear, story drifts etc. can be assessed correctly. Some code recommends the column dimension parallel to the longitudinal reinforcement of the beam should be at least 20 times the diameter of longitudinal bars for normal concrete. Few more general guidelines are followed for a beam element of a Specially Moment-Resisting Frame, lateral ties or stirrups and recommended to provide closely in the locations where flexural yielding is expected to occur, particularly at the end of the beams or the locations at lap/spliced bars. The transverse reinforcements are simultaneously act as confinement reinforcement to achieve more ductility of beams, columns and their joints. Transverse reinforcements at joints are provided to confine the joint core and to improve anchorage of the longitudinal reinforcement of both beams and columns. The maximum spacing of lateral ties/stirrups recommended not to exceed 75 mm as per IS 1893-Part I-2016.

Reinforced Concrete Braced frame is not very common vertical seismic resisting system but offers architects an alternative to walls and moment-resisting frames, often concealed within the walls of building cores braced frames to resist the seismic forces transferred from diaphragms. Braced frames resist lateral loads by the transfer of axial forces (tension or compression) through diagonal bracing members. These brace members transfer forces from roof to the foundation. Concentric frames have braces connecting at the ends of elements. The use of Reinforced Concrete bracing member has potential advantage than other bracing like higher stiffness and stability. The bracing systems is generally provided on periphery between the columns. It may be X-braced frames, V-braced frames etc. The most important parameters are base shear and storey displacement. It is observed that X-braced frames are more efficient and safe during earthquake shaking when compared with moment-resisting frames with V-brace members. Moment-resisting frame undergoes higher storey displacement compared to braced frame buildings, as both strength and stiffness of braced frame is much more compare to un-braced building frames. Base shear of braced frame buildings increases compared to building without brace member. Storey displacement of the building is reduced by to a great extent. By providing braces in the frame, the horizontal load at node is distributed among brace members along with beams and columns. Due to provision of the bracing system in the building, bending moment comparatively reduced.

Shear wall is a structural member in a reinforced concrete framed structure to resist lateral forces such as wind and earthquake forces. Shear walls provide large strength and stiffness to buildings in the direction of their orientation, which significantly reduces lateral sway of the building and as a results damage reduces. Moment-resisting frames carry lateral loads primarily by flexure. Joints are designed in such a way that they are completely rigid and therefore any lateral deflection of the frame occurs due to bending of the three-dimensional frame. Generally, shear walls are provided in highrise/slender buildings. Moment-resisting frames are more flexible than shear wall structures or braced frames but the horizontal deflection or drift, is greater. Adjacent buildings cannot be located too close to each other. But, in general, the energy dissipating capacity of Reinforced Concrete shear walls is not that good and it is found that using the bracing system may provide better solution. It is observed from experimental results that braced and infill bare frames has shown a higher lateral strength compare to bare frame depending on the type of bracing and infill. The energy dissipation for the braced and infill frames is always higher than that for the bare frame up to failure.

Potentially high ductile system with a higher degree of redundancy, allows freedom in architectural planning of internal spaces and external cladding. Its flexibility and associated long period may serve to detune the structure from the forcing motions on stiff soil or rock sites. However, poorly designed, moment-resisting frames have been observed to fail catastrophically in earthquakes, due to formation of weak stories and failure of beam–column joints. Beam–column joints are zones of high stress concentration. Although most of the shear walls are rectangular in plan. It may be a gentle curve or C, L and I-shapes which are usually structurally feasible. However, shear wall is effective in the direction of its length only. A structurally adequate wall possesses sufficient strength to resist both shear forces and bending moments. Cutouts for windows and doors may reduce the strength of a shear wall significantly. Therefore, such cutouts may be avoided at highly stressed zones, particularly near its base.

1.5 COLLAPSE PATTERN

Collapse due to unintended addition of stiffness, inadequate beam–column joint strength, tension & compression failures, wall-to-roof interconnection failures, local column failure, torsional effects, soft & weak story collapse, progressive collapse etc. are extremely important aspects and need to be discussed.

The following vertical irregularities repeatedly observed severe damages during earthquake shaking

- an abrupt change in the floor plan dimensions
- the columns on a particular floor level are more flexible and/or weaker than those above
- the short column effect
- discontinuous and offset structural walls
- a particular floor is significantly heavier than an adjacent floor

If some columns are shorter than others columns nearby, creates serious problem during earthquake shaking. If a particular column is half the height of the other columns nearby, the stiffness of the shorter column provides much higher stiffness than others. Stiffness is inversely proportional to the cube of column length (L^3). In this case, the shorter column is therefore eight times stiffer than the others as the length is half than others. Therefore, the shorter column theoretically will be able to resist eight times as much inertia force as the other columns, during earthquake shaking. It is unlikely to be strong enough to resist such a large inertia force and as a result may fail, lead instability of the building frame. This is commonly known as "Short column effect". Short column effect due to presence of unreinforced masonry infill of particular height (not full height), presence very deep spandrel beams, presence of a mezzanine slab, presence of stair beam/slab, presence of a plinth beam/tie beams, unequal basement columns on sloping ground.

Good ductility is achieved in a building when the collapse mechanism is of the desirable type. In such a case, the hysteretic loops of its load-deformation curve are stable and full. These type of hysteretic loops imply good energy dissipation in the building through each of the inelastic hinges at the beam ends. Such a behavior is observed in buildings that fail through sway mechanism, which ensures that beams yield before columns, and ductile flexural damages occur at beam-ends; this happens when the building has strong column – weak beam design. Beams are made to be weaker and ductile links. The buildings that fail through storey mechanism, damages are concentrated in the columns at a particular level and ductility demand on the columns is large. It arises, when the building has weak column – strong beam combination and the collapse mechanism dissipates less energy and that too also at a particular level. Hysteresis loop of a building structure depends on type of collapse mechanism.

Beam–column joints in moment-resisting frames need to be handled with care. Repairing of damaged joints is difficult. Beam–column joints must be designed and detailed to resist earthquake effects. Beams adjoining a joint are subjected to bending moments may be sagging or hogging in nature under cyclic earthquake shaking. Under this condition, the top bars in the beam-column joint are pulled in one direction on one side and pushed in the same direction on the other side and similar situation in the bottom bars also. Bond between concrete and steel plays an important role in the joint region. In such circumstances, there is a possibility of bond slip inside the joint region and beams lose their capacity to carry load. Further, under the action of the above pull-push forces at top and bottom ends, joints undergo distortion. Closely spaced closed-loop steel ties with 135° hooks are required around column bars to hold together concrete in joint region and to resist shear forces. Providing closed-loop ties in the joint requires some extra effort. Seismic design codes recommend continuing the transverse loops around the column bars through the joint region.

Strength discontinuity or sudden reduction in lateral strength along the height of the building, is a serious problem. This discontinuity or reduction causes large inelastic demand at the junctions. This kind of poor seismic structural configuration need to be avoids if possible. Inelastic pushover analysis may be helpful after designing the building for different load combinations. Is 1893 –Part I – 2016, recommended in several way to avoid this kind of problems. The distribution of lateral strength and stiffness along the building height needs to be determined to identify any irregularity in lateral strength or stiffness.

Columns sizes have to be larger to be able to ensure that they do not fail in a brittle manner. There are a number of joints, which may be inefficient in transferring the forces and moments, if not designed, detailed and constructed properly. IS 13920 recommended the details in the design and detailing of Reinforced Concrete beams and columns as well as beam–column joints. Shear walls, being stiffer than moment frames, attract more earthquake force toward themselves. This facilitates moment frames to be lightly reinforced, which may lead to more economical solution. The effectiveness of Reinforced Concrete structural walls, need to utilized with a pinpointed objectives. Open ground storey RC frame buildings are common in throughout the world, knowing poor performances of such buildings. There is no way out but to avoid different type of irregularity. IS 1893 (Part 1), 2016, suggests that the strength of any storey should not be less than 80 percent of that in the storey above.

Two parts of the same building are built close to each other and are separated by small gap for construction joint. Pounding between two parts can be avoided by calculating need for actual separation required to avoid pounding between the parts during seismic shaking. This is also applicable for closely spaced tall buildings where tip displacement will be very high and possibility of pounding between two building cannot be ignored.

Soft storey problem arises during earthquake shaking, if a particular storey of a building is more flexible or weaker than the storey above it. The columns of that soft storey will be seriously damaged. This problem may cause more damage due to questionable vertical configuration. Open ground floors allowing relatively more risk of a soft storey mechanism of the buildings located in a highly seismic zone. Soft storey risk will be more, where a combination of open ground floors and masonry infill frame, exist. Structural members with additional strength and ductile, are required to tackle this problem. A building having lateral strength is less than that of storey above, then it is called weak storey. Generally, when seismic weight of any floor is more than 1.5 times of the floor below, then the building behavior changes due to mass irregularity (refer to Figure 1.5 and 1.6).

1.6 DYNAMIC RESPONSE CONCEPT

Important dynamic characteristics of buildings are natural periods, mode shapes and damping. Each of these natural frequencies, and the associated mode shape of a building, contribute to displacement of the structure. Three fundamental

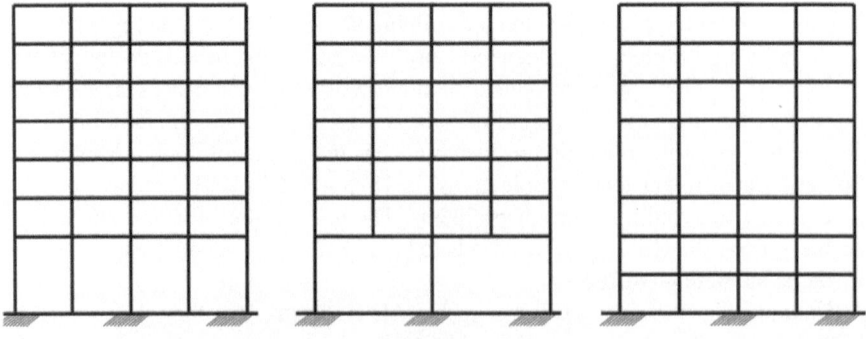

ELEVATION
IRREGULAR FRAME WITH SOFT AND WEAK STOREY

FIGURE 1.5   Strength irregularity

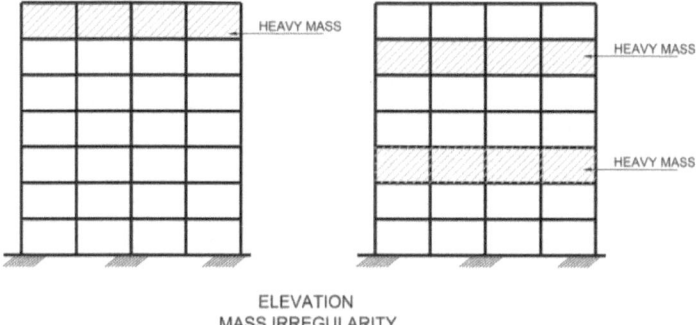

ELEVATION
MASS IRREGULARITY

FIGURE 1.6   Mass irregularity

translational natural periods, T_x, T_y and T_z, are associated with its horizontal translational oscillation along the X and Y directions and vertical translational oscillation along the Z direction respectively and one fundamental rotational natural period, $T_{\theta 1}$, is associated with its rotation about an axis parallel to the Z axis. For a building, there are numbers of natural periods. Each node is free to translate in all three Cartesian directions and rotate about the three Cartesian axes. Hence, if the number of nodes is N, then there would be 6N number of mode shapes and associated with 6N number of natural periods. Irregular buildings which have irregular geometry, non-uniform distribution of mass and stiffness in plan and along the height, have mixed mode shapes. The overall response of a building is the sum of the responses of all of its modes. The contributions of different modes shapes vary, usually, contributions of some modes dominate. It is important to endeavor to make buildings regular to the extent possible. But, in a regular buildings too, care should be taken to locate and size the structural elements such that torsional and mixed modes of oscillation do not participate much in the overall oscillatory motion of the building. One way of avoiding torsional modes to be the early modes of oscillation in buildings is increasing the torsional stiffness of building. This is achieved by adding in-plane stiffness in the vertical plane in select bays along the perimeter of the building, this addition of stiffness should be done along both plan directions of the building, such that the building has no stiffness eccentricity. Adding braces or introducing structural walls in selected bays are common ways. Also, there are a number of possibilities in which buildings can oscillate along each direction of oscillation. The mode shapes of buildings depend on the overall geometry of the building, the geometric and material properties of its structural members, and connections between the structural members and the soil below the building. Buildings exhibit flexural mode shape, shear mode shape or a combination of these depending on the above factors.

The natural period of a building is the time taken by it to undergo one complete cycle of oscillation. It is an inherent property of a building, controlled by its elemental mass and stiffness as well as the mass and stiffness distribution. Buildings that are heavy, with a larger mass, and flexible have a longer natural period than light and stiff buildings. The reciprocal of the natural period of a building is called the natural frequency; its unit is Hertz (Hz). Increasing the column size increases both the stiffness and the mass of buildings. But, when the percentage increase in stiffness as a result of an increase in the column size is greater than the percentage increase in mass, the natural period reduces. But, increase in column size reduces the natural period of buildings, does not consider the simultaneous increase in mass, in that context, buildings are said to have shorter natural periods with increase in column size. Proper estimation of flexural stiffness of each individual members is essential for predicting dynamic characteristics of a building as well as force and deformation. Reinforced concrete poses a special challenge of predicting the most suitable cross-section property, when sections undergo extensive cracking during earthquake shaking. The choice is between gross and cracked cross-sectional properties associated with axial, flexural, shear and torsional actions. Gross cross-sectional properties are computed using gross sectional area without considering the stiffness enhancement due to the presence of longitudinal reinforcement and the extent of cracking of the member is assumed to be minimal. Often, gross properties are commonly used for estimating force and deformation demands on members subjected to gravity loading based on linear analysis. But, in members where extensive cracking is expected during earthquake shaking, estimation of force and deformation demands based on gross properties may not provide true behavior. Effective properties are necessary to overcome this shortcomings and represent reduced stiffness of members in their damaged state. Effective properties are established based on extensive analytical and experimental studies on buildings/members subjected to seismic loading and expressed as a fraction of gross stiffness. For instance, the ratio of effective moment of inertia to gross moment of inertia of columns is higher than that of beams, because damage expected in columns is lower owing to presence of compressive axial load on it. The actual ratio depends, for example, on the level of compressive axial load, among many other factors; thus, literature on the subject has different suggestions. IS 1893, Part I, 2016 suggests that $I_{b,eff} = 0.3 \ XI_{b,gross}$ for beams and $I_{c,eff} = 0.70 \ I_{c,gross}$ for columns. Using these values of effective moment of inertia, the fundamental natural periods of buildings, is to be calculated.

1.7 WIND LOAD AND EARTHQUAKE LOAD

1.7.1 Wind Load

The effect of wind is more crucial for tall buildings. Innovations in architectural form and advances in structural analysis have made highrise buildings lighter and more prone to sway. Wind load is dynamic in nature and its impact on structure depends on its time period, randomness, shape of the building on plan, irregularities etc. Wind having long periods of time, may be considered as more or less static. However, wind having shorter time periods need to be considered as dynamic one. The dynamic wind pressure produces narrow-band random vibration of the building in both along and across wind as well as rotational wind. The magnitudes of the sway depends on the distribution of wind velocity and direction as well as on the mass, stiffness and shape of the building. In many cases the across-wind effect is more severe than the along-wind effect. However, the response of the building need to be obtained with care for a relatively flexible building. In general, dominant mode shapes is extremely important to predict dynamic response. The stiffness of the building affects the resonance frequencies. If the building is stiff, the resonance frequencies will be relatively high and the dynamic deflections will not be significant. Important and relevant design parameter need to be considered to arrive at design wind pressure and its distribution. In this regard, lifetime of the building is of immense importance. Generally, return period is considered equal to the lifetime of the building. For most of the small buildings, "Equivalent static approach" may be applied. However, if the building is flexible and the resonance frequencies will become lower, where frequency of the fluctuating wind forces, may be critical. The building will tend to be influenced by the fluctuating wind actions below the fundamental frequency and will be attenuated at frequencies above. Deflections will be extremely high, if resonance occurs. For flexible buildings the oscillations may interact with the aerodynamic forces which may lead to instability in form of galloping oscillations, vortex shedding, divergence, flutter etc. When a building is excited by wind load, the wind force tend to be random in amplitude and spread over a wide range of frequencies. The response of the building is then decided by the wind energy available in the narrow bands close to the natural frequencies of the building. The major part of the exciting wind energy often occurs at frequencies lower than the fundamental frequency of the building and the energy decreases with increasing frequency. The response to the along and across wind arises from wind–structure interaction mechanics where the along wind is primarily to buffeting effects caused by turbulence and the across wind is primarily to vortex shedding. The across-wind response is of special interest. IS 875, part III, 2015 recommends a number of concepts of wind assessment.

1.7.2 EARTHQUAKE LOAD

Buildings oscillate during earthquake shaking and inertia forces are induced at different levels of the building. The intensity and duration of oscillation, and the amount of inertia force induced in a building depend on dynamic characteristics of the building as well as the dynamic characteristics of the earthquake shaking. If the frequency of ground shaking is close to any of the natural frequencies of building, resonance will occur and the building will undergo a huge displacement and finally lead to collapse. The ground motion contains a basket of frequencies that are continually and randomly changing at each instant of time. There is no guarantee that the ground shaking contains the same frequency close to natural frequency of the building. Even the ground shaking occurs at frequencies close to natural frequency of the building, for a small period of time, building will undergo huge displacement lead to severe damage. It has been observed that, response increases, but resonance condition is not always reached. Mass of a building that is effective in lateral oscillation during earthquake shaking is called the seismic mass of the building. It is the sum of its seismic masses at different floor levels. Seismic mass at each floor level is equal to full dead load plus appropriate percentage of design live load, which depends on design live load intensity. Seismic design codes recommended this percentage of live loads to be considered on account of seismic mass. An increase in mass of a building increases its natural period. Heavier buildings have larger natural period. As the height of building increases, its mass increases but its overall stiffness decreases. Hence, the natural period of a building increases with increase in height. Orientation of rectangular columns influences lateral stiffness of buildings and period of building changes. Again, longer side of the rectangular column is generally oriented to the shorter direction of the building (i.e., weaker direction of the building) to economize the column section but this will make the building stiffer in shorter direction and as a result frequency will be more in shorter direction, i.e., time period will be less. If time period reduces base seismic shear will increase and push the bending moment and shear force on the higher side, which will lead to increase of column size and reinforcement and the cost will increase. Therefore, the size and orientation of column need to be finally decided after few trials, to achieve an optimal solution.

2 Wind Analysis of Buildings

2.1 PREAMBLE

Earth's rotation and differences in terrestrial radiation, are the important causes of wind. The speed of wind varies along height basically due to ground roughness. Generally, the horizontal component of wind is predominant though in many situation vertical component is also considered. Generally, the wind speed is measured by Anemometers. Wind speed increases with height and becomes maximum at gradient height. The variation with height depends primarily on the terrain conditions. In all terrains/ground roughness, wind speeds beyond gradient heights are equal. The magnitude of wind speed depends on the averaging time, at any height in a given terrain. Shorter the averaging time, the higher is the mean wind speed. Wind to travel over a typical terrain, takes a distance known as fetch length, to fully develop the velocity profile for that terrain category

The magnitude of fluctuating component, which represents the gustiness of wind, depends on the averaging time. The fluctuating velocity is normally expressed in terms of turbulence intensity. Time period of oscillation and damping of a flexible structure affects its response due to the gustiness or turbulence in wind. Wind characteristics get modified due to presence of obstructions. There may be interference effect with the structures in the close vicinity of the building. Internal pressure of the building increases due to external openings could result in suction or compressive on roof/walls Therefore, design values is the algebraic sum of the external and internal wind pressure. The IS code specifies the basic wind speed considering a gust of 3 second duration, i.e., the wind speed averaged over a 3-second period of time. Short period gusts may not cause any appreciable increase in pressure on the walls, roof etc. Effect of gusts may be severe for buildings with high slenderness ratios. Tributary area is also need to be accounted for. Reduction in wind pressures is specified for tributary area beyond 100 m^2 in different codes of practices. Tornados, whose effect is much more than the severest cyclones is a narrowband phenomenon of limited time duration, occurs in parts of the world and it is not yet could be assessed correctly. Wind is not a steady phenomenon because of its natural turbulence and gustiness. Wind causes a random time-dependent load, which has a mean component along with a fluctuating component. The building will undergo dynamic oscillation due to mostly due to this fluctuating component of wind. The oscillation is insignificant for a short rigid buildings. Therefore, in such situation, an equivalent static concept is generally used to assess design pressure and recommended in different codes of practices globally. A building is generally, considered under this category, if its natural time period is less than one second. This codes of practices have given guidelines for buildings during erection/construction and need to be followed during various stages of erection/construction. Strong wind along with icing, may occur simultaneously in different parts of the world. Values of wind pressure coefficients may be slightly different in different codes of practices. However, both pressure and force coefficients are derived on the basis of wind tunnel studies. The nature and magnitude of these wind parameters depends on primarily on, the geometry, point of separation, nature of the incident wind, direction of wind incidence, etc., i.e., basically characteristics of wind flow over or around a building.. Pressure coefficients are commonly based on the quasi–steady assumption, whereby the pressure coefficient is taken to be the ratio of mean pressure measured over a point or pressure averaged over a small tributary area divided by the dynamic pressure for the mean speed of incident wind. This approach implicitly assumes that the fluctuations in pressure follow directly the speed. Wind turbulence gets modified as it approaches the building, due to vortex shedding, etc. Influence of wind incidence angle particularly on edges and corners, is one of the important parameters. Recent versions of some international Codes have been revised considering different aspects of wind so far. Rectangular buildings with length to width ratio, buildings of different other shapes, both with sharp edges as well as rounded corners, are taken into consideration, in wind assessment by different codes of practices. Force coefficients, which imply considerations of their shape, aspect ratio, Reynolds number and shielding.

Generally, for tall flexible buildings, when time period of the building is greater than one second, would respond to wind dynamically not only in the direction of wind but also transverse direction to the wind flow Interference effects are primarily due to modifications in the incident and wake flow characteristics. Generally, the dynamic part of the wind pressures would set up oscillations in a flexible building, if fundamental time period is more than one second. Furthermore, flexible buildings also respond in the across-wind direction due to vortex shedding. Interfering structures, vortex shedding frequency may come close to the natural frequency of the building. The values may vary widely depending upon the distance and height of the buildings involved. These complex aspects of interference need to be studied in the wind tunnel. A flexible building would tend to oscillate due to vortex shedding alternately from either sides of the building at regular intervals, in the cross-wind direction, causing a dynamic force. The frequency of vortex shedding is dependent on dimension of the building, shape and wind speed, height, etc. Vortex shedding forces acting in a direction normal to the direction of wind causing across-wind as well as torsional response. A non-dimensional parameter named as "Strouhal number", to take care of periodicity of vortex shedding, is introduced in different wind assessment. It depends on shape of the building also.

Different codes suggested procedures for crosswind assessment. For tall buildings and, the contribution from higher modes of vibration being rarely significant. The magnitude of the across-wind force and the pitching moment thus produced would depend not only on the turbulence level but also the mean wind speed and the angle of attack. Computation of these forces need wind tunnel studies as well as CFD (computational fluid dynamics) analysis

2.2   WIND LOAD PROVISIONS AS PER IS 875 (PART 3), 2015

The wind map as per IS 875 Part 3, 2015 have to be used to have wind speed in a particular place

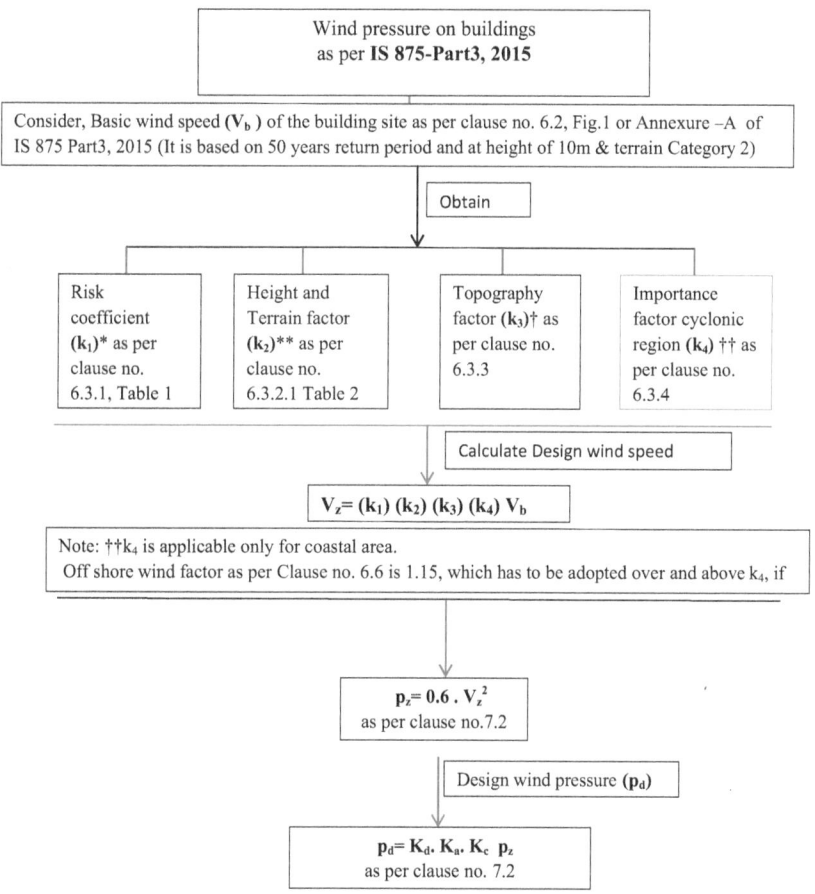

Flow chart to calculate wind pressure on buildings

k_d = Wind directionality factor (as per clause 7.2.1). The wind pressure or force coefficients are determined for a specific wind direction, but in fact the wind directionality on a structure is random, for which a multiplication factor, k_d, is to be adopted. A factor of 0.90 may be used for the design wind pressure. For circular or near-circular forms, this factor may be taken as 1.0. For cyclone-affected regions as well, the factor k_d is to be taken as 1.0.

k_a = Area averaging factor (clause 7.2.2), table 4 (amendment 1). The pressure coefficients adopted to calculate forces are a result of averaging the measured pressure values over a given area. For a small area the value matches, but for a larger area the correlation of the measured values decreases, for which a multiplication factor, k_a, is to be adopted.

k_c = Combination factor frame (as per clause 7.3.3.13).To calculate the wind loads on the frames of clad buildings, the pressure or suction inside the structure and the pressure or suction outside the structure are not taken to be fully correlated. So, for calculating the combined effect of the wind loads on frames, a multiplication factor, k_c, has to be adopted.

Notes:

* Factor k_1 is based on statistical concepts that take into account the degree of reliability and the return period, which is generally considered equal to the life of the structure.

** Basic wind speed calculated at 10 m height and terrain category 2. For other terrain categories and heights, a multiplication factor, k_2, to be adopted.

† Basic wind speed calculated considering a flat site above sea level but it does not incorporate the effect of topographical features of site like hills, valleys, cliffs etc., for which wind speed may increase or decrease. To consider this topographical effect k_3 factor need to be adopted.

†† In coastal area (up to 60 km from coast) due to severe cyclonic effect, basic wind speed exceeds a given value for non-coastal areas. k_4 factor need to be adopted for coastal area.

As per clause 6.3.3.1 of IS 875 (Part 3), 2015, the topography factor k_3 = 1.0, taking the upwind slope to be less than 3^0. If the upwind slope is more than 3^0, the value of k_3 needs to be calculated as per annexure C of IS 875 (Part 3), 2015.
As per clause 7.2.1 of IS 875 (Part 3), 2015:

$$\textbf{Design wind pressure } \mathbf{p_d = K_d \cdot K_a \cdot K_c \cdot p_z}$$

(but p_d is not to be less than $0.7p_z$)

2.2.1 Different Approaches to Wind Analysis

Now, wind–structure interaction effects are considered to have a design load on structures.

IS 875 (Part 3), 2015, suggests three different concepts to arrive at the design loads. These concepts are as follows:

- the pressure coefficient approach
- the drag coefficient approach
- the gust factor approach

2.2.1.1 Pressure Coefficient Approach

As per clause 7.1, the wind load on a building may be calculated for individual structural elements, such as the roof and walls.

As per clause 7.3.1 of IS 875 (Part 3), 2015:

Design wind force (F) = (Cpe–Cpi) A. p_d

where

C_{pe} = external pressure coefficient
C_{pi} = internal pressure coefficient
A = surface area of structural element or cladding unit
p_d = design wind pressure

* Pressure coefficient as per clause 7.3 of IS 875 (Part 3), 2015, states: "The pressure coefficients are always given for a particular surface or part of the surface of a building. The wind load acting normal to a surface is obtained by multiplying the area of that surface or its appropriate portion by the pressure coefficient (C_p) and the design wind pressure at the height of the surface from the ground."

Note:

1. If the surface design pressure varies with height, the surface area of the structural element may be subdivided, so that the specified pressures are taken over appropriate areas.
2. Positive wind pressure means that the pressure is acting toward the structure and negative means away from it.

The internal pressure coefficient depends upon the degree of permeability of the building, which depends primarily on the percentage of the external wall surface that consists of openings.

As per clause 7.3.2.1, in the case of buildings where the claddings permit the flow of air with openings not more than about **5 percent** of the wall area but where there are no large openings, it is necessary to consider the possibility of the internal pressure being positive or negative. Two design conditions shall be examined, one with an internal pressure coefficient of **+0.2** and another with an internal pressure coefficient of **-0.2**.

Clause 7.3.2.2 deals with medium to large openings:

Buildings with medium and large openings may also exhibit either positive or negative internal pressure depending upon the direction of wind. Buildings with medium openings between about **5 and 20 percent** of wall area shall be examined for an internal pressure coefficient of **+0.5** and later with an internal pressure coefficient of **-0.5**, and the analysis which produces greater distress of the member shall be adopted. Buildings with large openings, that is, openings **larger than 20 percent** of the wall area shall be examined once with an internal pressure coefficient of **+0.7** and again with an internal pressure coefficient of **-0.7**, and the analysis which produces greater distress of the member shall be adopted.

So, internal pressure coefficient need to be adopted considering the percentage of opening of wall of the building.

Consider individual structural elements of the building like walls.

As per clause 7.3.1 of IS 875 (Part 3), 2015:

$$F = (C_{pe} - C_{pi}).\,A.\,p_d$$

where

F = wind force acting at a particular point
A = frontal area
C_{pe} = external pressure coefficient
C_{pi} = internal pressure coefficient
p_d = design wind pressure

Numerical examples are available in Chapter 4.

2.2.1.2   Drag Coefficient Approach

Considering the building as a whole and closed. It is considered as a vertical cantilever fixed at foundation level. The Drag force coefficients (C_f) shall be obtained as per is to be obtained as per clause no. 7.4

Clause 7.4 states: "The value of force coefficients (C_f) apply to a building or structure as a whole, and when multiplied by the effective frontal area (A_e) of the building or structure and design wind pressure, (p_d) gives the total wind load (F) on that particular building or structure."

$$F = C_f.\,A_e.\,p_d$$

where

F = wind force acting in a specified direction on the frontal area
C_f = force coefficient on the building (as per clause 7.4.2)
A_e = effective frontal area of the building
P_d = design wind pressure

The force coefficients for different rectangular-shaped buildings are given in figures 4(a) and 4(b) of IS 875 (Part 3), 2015, depending on the plan ratio (a/b) and the height/breadth (h/b)ratio.

C_f can be obtained from table 25, depending on the plan ratio (a/b), the height/breadth ratio (h/b), the wind speed/width ratio and the surface roughness. The higher value obtained from table 25 and figure 4 of IS 875 (Part 3), 2015, has to be used for calculating the design wind force acting on the structure.

2.2.1.3   Gust Factor Approach

As per clause 9.1 of IS 875 (Part 3), 2015, the dynamic effects of the wind have to be considered, and the wind-induced oscillation effect has to be examined, if

(i) h/b > 5

where

> h = height of the building
> b = width of the building

and/or

ii)   Natural frequency of the building in first mode < 1 Hz (cycles/second)

(again, as per clause 9.2.1, note 4)
The vortex shedding effect needs to be considered if

l/b < 2

where

> l = length of the building
> b = width of the building

Calculation of the wind force proceeds as follows.

Along-Wind Response

$$F_z = C_{f,z.} \, A_{z.} p_{d.} G$$

where

> F_z = design peak along-wind load on the building structure at any height z
> $C_{f,z}$ = drag force coefficient of the building structure corresponding to the area A_z
> p_d = design hourly mean wind pressure corresponding to V_{zd} and obtained as $0.6 \, V_{zd}^2$ (N/m²)
> V_{zd} = design hourly mean wind speed at height z, in m/s
> A_z = effective frontal area of the building structure at any height z, in m²
> G = gust factor: $1 + r \, \sqrt{[g_v^2 B_s (1 + \varnothing)^2 + H_s g_r^2 SE/\beta]}$

where

> r = roughness factor, which is twice the longitudinal turbulence intensity, $I_{h,i}$
> g_v = peak factor for upwind velocity fluctuation (3.0 for category 1 and 2 terrains, 4.0 for category 3 and 4 terrains)

B_s = background factor, indicating the measure of the slowly varying component of the fluctuating wind load caused by the lower-frequency wind speed variations: $B_s = 1/[1+ \sqrt{(0.26(h-s)^2 + 0.46\, b_{sh}^2 / L_h)}]$

where

b_{sh} = average breadth of the building/structure between heights s and h
L_h = measure of effective turbulence length scale at the height, h, in m ($= 85\, [h/10]^{0.25}$ for terrain category 1 to 3, $= 70\, [h/10]^{0.25}$ for terrain category 4)
$\emptyset$ = factor to account for the second-order turbulence intensity: $= g_v I_{h,i} \sqrt{B_s}/2$

where

$I_{h,i}$ = turbulence intensity at height h in terrain category 1
H_s = height factor for resonance response: $= 1 + (s/h)^2$

where

s = size reduction factor, given by $= 1/[1 + (3.5 f_a\, h)/V_{hd}][1 + (4 f_a b_{oh})/V_{hd}]$

where

b_{oh} = average breadth of the building/structure between 0 and h
E = spectrum of turbulence in the approaching wind stream: $= \pi N/ (1 + 70.8\, N^2)^{5/6}$

where

N = effective reduced frequency $= f_a L_h/V_{hd}$

where

f_a = first mode natural frequency of the building/structure in along-wind direction, in Hz
V_{hd} = design hourly mean wind speed at height, h in m/s
ß = damping coefficient of the building/structure (for Reinforced Concrete structures, it is 0.020)
g_r = Peak factor for resonant response $= \sqrt{2\ln(3600 f_a)}$

Hourly mean wind speed

$$V_z H = k_{21} Vb$$

As per clause 6.4:
k_{21} = hourly mean wind speed factor for terrain category 1: $= 0.1423[\ln(z/z_{0i})]$ $(z_{0i})^{0.0706}$

where

z = height or distance above the ground
z_{0i} = aerodynamic roughness height for i^{th} terrain

The design hourly mean wind speed

$$V_{zd} = V_b \, k_1 \, k_2 \, k_3 \, k_4$$

Across-Wind Response

As per clause 10.3 of IS 875 (Part 3), 2015, the across-wind load on the building will be a distributed one
The distribution of loads is as follows.

$$F_{z,c} = (3M_c / h^2) \, (z/h)$$

where

$F_{z,c}$ = across-wind load per unit height at height z
$M_c = 0.5 \, g_h \, p_h \, b \, h^2 \, (1.06 - 0.06k) \, \sqrt{(\pi C_{fz}/\beta)}$

where

g_r = peak factor = $2 \sqrt{(2 \ln(3600 \, f_c))}$
p_h = hourly mean wind pressure at height h, in Pa
b = breadth of the structure normal to the wind, in meters
h = height of the structure, in meters
k = mode shape power exponent for representation of the fundamental mode shape, as represented by $-\psi \, (z) = (z/h)^k$
f_c = first mode natural frequency of the building/structure in the across-wind direction, in Hz

The cross-wind force spectrum coefficient (C_{fs}) can be calculated from Figure 11 in IS 875 (Part 3), 2015. From Figure 11 (for rectangular building), C_{fs} is to be calculated Turbulence intensity of 0.2 at 2/3 h curve has been considered.

k = mode shape power exponent for slender-framed structure (moment-resisting) = 0.5
β = damping coefficient of the building/structure for Reinforced Concrete structure = 0.02

A numerical example has been done in Chapter 4.

3 Seismic Analysis of Buildings

3.1 PREAMBLE

According to building codes, earthquake-resistant structures are intended to withstand the largest earthquake of a certain probability that is likely to occur at the building's location. This means that the loss of life should be minimized by preventing the collapse of the building in the event of rare earthquakes while the loss of the functionality should be limited for more frequent ones.

Currently there are several design philosophies in connection with earthquake engineering, making use of experimental results, computer simulations and observations from past earthquakes to offer the required performance in terms of assessing the seismic threat at the site of interest. These range from appropriately sizing the structure, so as to be strong and ductile enough to survive the shaking with an acceptable level of damage, to equipping it with base isolation or using structural vibration control technologies to minimize any forces and deformations. The capacity design concept requires a hierarchy of structural component strengths in the design of structures, aiming to ensure that inelasticity is confined to predetermined and preferred structural components.

3.2 SEISMICITY

Any sudden shaking of the ground is caused by the passage of seismic waves. Seismic waves are produced when some form of energy stored in Earth's crust is suddenly released. Generally it happens when masses of rock straining against one another suddenly fracture, and slip occurs. Earthquakes occur along geologic faults–narrow zones where rock masses are moving in relation to one another. The major fault lines are located at the fringes of tectonic plates. Tectonic earthquakes are explained by the elastic rebound theory, according to which theory a tectonic earthquake occurs when strains in rock masses have accumulated to a point where the resulting stresses exceed the strength of the rocks, and sudden fracturing occurs. The fractures propagate rapidly through the rock, usually tending in the same direction and sometimes extending many kilometers along a local zone of weakness. As the fault rupture progresses along or up the fault, rock masses are flung in opposite directions, and thus spring back to a position where there is less strain. At any one point this movement

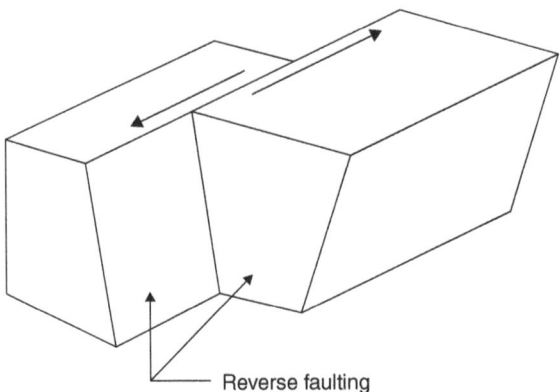

Reverse faulting

FIGURE 3.1   Reverse faulting

may take place not at once but, rather, in irregular steps, giving rise to the vibrations that propagate as seismic waves. Fault rupture starts at the earthquake focus, a spot that in many cases may be even 10–15 km below the surface of the Earth. The rupture propagates in one or both directions over the fault plane until it is stopped or slowed. Sometimes, instead of being stopped, the fault rupture recommences on the far side and the rupture continues.

Earthquakes have different effects depending on the type of fault slip. The usual fault has a "strike" – i.e., the direction from north taken by a horizontal line in the fault plane – and a "dip": the angle from the horizontal shown by the steepest slope in the fault. The lower wall of an inclined fault is called the footwall. Lying over the footwall is the hanging wall. When rock masses slip past each other parallel to the strike, the movement is known as strike-slip faulting. Movement parallel to the dip is called dip-slip faulting. Strike-slip faults are right lateral or left lateral, depending on whether the block on the opposite side of the fault from an observer has moved to the right or left. In dip-slip faults, if the hanging wall block moves downward relative to the footwall block, it is called "normal" faulting; the opposite motion, with the hanging wall moving upward relative to the footwall, produces "reverse" faulting, as shown in Figures 3.1 and 3.2.

All known faults are assumed to have been the seat of one or more earthquakes in the past, though tectonic movements along faults are often slow, and most geologic-ally ancient faults are now aseismic – i.e., they no longer cause earthquakes. The actual faulting associated with an earthquake may be complex, and it is often not clear whether in a particular earthquake the total energy issues from a single fault plane.

MSK INTENSITY SCALE

A number of different magnitude scales are used by scientists and engineers as a measure of earthquake intensity. The P-wave magnitude (M_b) is defined in terms of the amplitude of the P wave recorded on a standard seismograph. Generally, the surface-wave magnitude (M_s) is defined in terms of the logarithm of the maximum

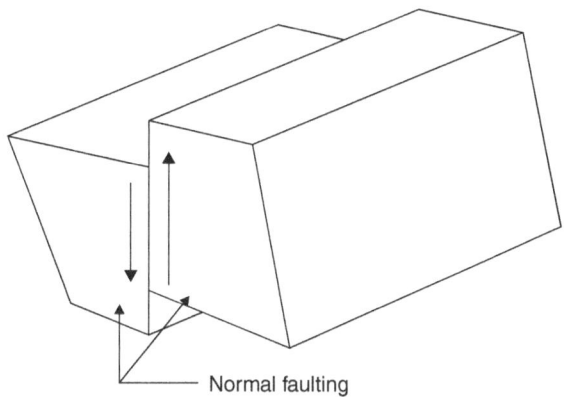

Normal faulting

FIGURE 3.2 Normal faulting

amplitude of ground motion for surface waves with a wave period of 20 seconds. Sensitive seismographs can record earthquakes with magnitudes of negative value and up to about 9.0.

A mechanical measure of earthquake size is also used, namely the seismic moment (M_0). Such a parameter is related to the angular leverage of the forces that produce the slip on the causative fault. It can be calculated both from recorded seismic waves and from field measurements of the size of the fault rupture. Consequently, the seismic moment provides a more uniform scale of earthquake size based on classical mechanics. It allows a more scientific magnitude to be used, known as moment magnitude (M_w). It is proportional to the logarithm of the seismic moment; values do not differ to a great extent from M_s values. For example, the Alaska earthquake of 1964, with a Richter magnitude (M_L) of 8.3, also had the values $M_s = 8.4$, $M_0 = 820 \times 10^{27}$ dyne centimeters and $M_w = 9.2$. The moment magnitude scale was designed to produce a more accurate accounting of the total energy released. It calculates the magnitude of earthquake more accurately than other way of measuring earthquake intensity – i.e., Richter scale (M_L), body-wave scale (m_b) and surface-wave scale (M_s). The moment magnitude scale considers the fault's geometry, the angle and other qualities of the plane that characterize the fault that ruptures during an earthquake and seismic moment – that is, the displacement of the fault across its entire surface multiplied by the force to move the fault. Seismographs are also used to provide the data used to calculate the seismic moment. Instead of relying only on the peak amplitude of the largest incoming seismic wave (as in the Richter scale), measurements taken from seismographs at different locations are used to describe seismic waves emanating from the focus. The seismic moment, M_0, can be expressed by the formula $M_0 = DA\mu$, where D is the average fault displacement, A is the total area of the fault surface and μ is the average rigidity with respect to the shearing forces of the rocks in the fault. M_0, measured in dyne cm, is, essentially, the amount of energy released during an earthquake. Moment magnitude (M_w) can be determined by using the following formula:

$$M_w = 2/3 \log M_0 - 10.7$$

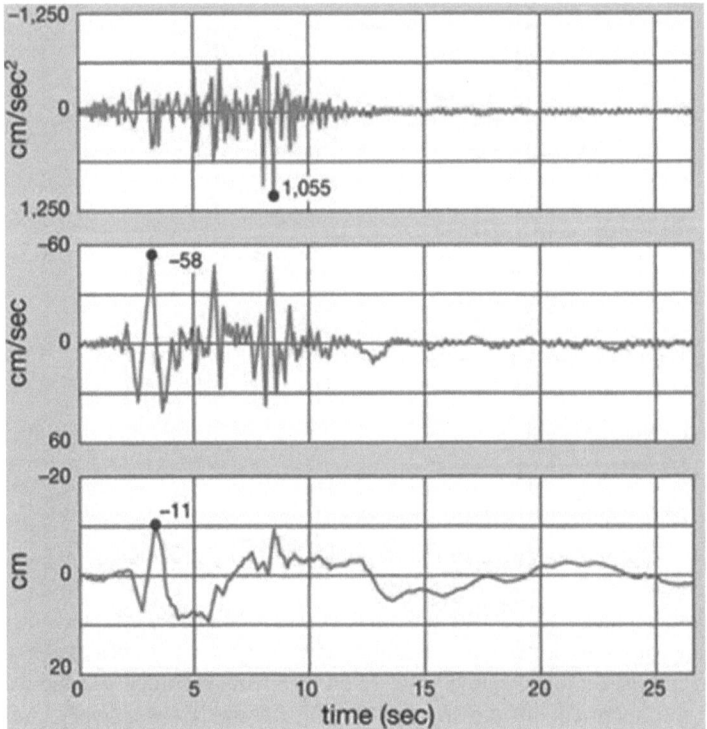

FIGURE 3.3   A seismograph

The seismograph is a graphical record that is subsequently analyzed. A typical seismograph is shown in Figure 3.3.

3.3   GENERAL PRINCIPLES AND DESIGN CRITERIA

General principles and design criteria of earthquake-resistant design of buildings need to be discussed for better understanding between structural engineer, architect and owner. The major aspects are structural configuration, lateral stiffness, lateral strength and ductility aesthetics, and functionality of the building are to be consider in parallel. Shape and size of the building, location and size of structural and non-structural elements, are the major design aspects. It is a quite well known fact that convex geometry of building in elevation is preferred than that a concave geometry, for smooth transfer of load from top to foundation level. Convex shaped buildings have direct load paths for transferring earthquake shaking induced inertia forces to the base level for any direction of ground shaking, avoiding stress concentration problems. No doubt that buildings with rectangular plans and straight elevation is the best solution for earthquake resistant buildings, because inertia forces are transferred without having to bend due to the geometry of the building. Buildings with setbacks and central openings offer geometric constraint to the flow of inertia forces. If the inertia force paths have to bend before reaching the ground are normally not preferred. "Lateral stiffness" means the initial stiffness of the building without damage. "Lateral strength" means the maximum resistance. Ductility toward lateral deformation is measured in terms of a ratio of the maximum deformation and the yield deformation. Earthquake imposes displacement at the foundation level of the building, which is time varying and random in nature. It demands lateral deformation in the building, i.e., its sub-structure to super-structure. The building should be able to withstand deformation with damage under small intensity shaking, and no collapse under high intensity shaking. The building needs to have large inelastic deformation capacity and needs to have the sufficient strength in all its members, to sustain the forces and moments induced in them, during earthquake shaking. Deformation demand on the building depends on the seismic condition of the building site and the deformation capacity of the building can be controlled by architects and structural engineers. It is not really possible to predict maximum possible ground displacement at a building site by Geotechnical Engineers / Earth scientists. It is required to understand overall nonlinear behavior of a Reinforced Concrete building to predict ultimate deformation capacity. It can be done to a great extent due to availability of computer software which can take care of complex mathematical analysis.

Hierarchy of structural component strengths is the basic requirement of capacity design concept, in order to ensure pre-decided failure mechanism. Non-ductile failure modes are delayed by providing higher resistance to this kind of modes of failure. The load path in a moment resisting Reinforced Concrete frame building involves slab and beam elements, beam–column joints, column element and foundations, including soil below the foundation. Strength hierarchy is essential along the load transfer path. Stronger structural elements are required which are supporting other structural elements in sequence. For example columns should be stronger than primary frame beams. Again, primary frame beams should be stronger than secondary beams. The beam–column joints should be strong enough not only to hold the beam and columns but should be able to contribute to overall ductility of the building. Therefore, beam–column joints should have enough ductile. No plastic hinges should be formed in columns in order to ensure sway mechanism of failure only. It may be noted here that it is not easy to strengthen the damaged beam-column joint, after earthquake. Beams should be stronger than adjoining braces (if provided). Columns should be stronger than adjoining beams. Beam-column joint should be stronger than the adjoining columns and beams. Foundations should be stronger than associated columns. Finally, soil below foundations, should be strong enough to withstand loads during earthquake shaking. If required, appropriate soil strengthening, need to be done.

During an earthquake shaking, the building is subjected to random motion of the ground at its base. This, in turn, induces inertia forces in building frame at different levels along the height. Usually beams, are designed against bending moment and shear force independently. Both the possible failure modes (flexure and shear) are precluded independently through design. The beam should be ductile also. For a under-reinforced flexural action, mode of failure precedes the non-ductile or brittle mode of failure due to shear.

3.4   RESPONSE SPECTRUM OF A GROUND MOTION

3.4.1   ACCELERATION RESPONSE SPECTRUM OF A GROUND MOTION

The maximum force induced during earthquake shaking, is necessary for designing a building.

Further, since absolute maximum of such response is useful in design, a graph of the maximum response need to be generated for a spectrum of single degree of freedom structures with different natural periods keeping damping constant under an earthquake ground motion. This graph is called as the "response spectrum graph" under a particular earthquake ground motion. One such response spectrum graph corresponding to the acceleration of the building is called as the *acceleration response spectrum*". IS 1893-Part I-2016, recommended acceleration response spectrums for different types of soil. The mass of the building which is effective during earthquake shaking, called as seismic mass/seismic weight. Thus, once the natural period associated with each mode of oscillation is estimated, the corresponding "*design seismic acceleration coefficient* (s_a/g)", can be obtained, considering appropriate type of soil and damping percentage of reinforced concrete (generally considered 5%). The generation of acceleration response spectrum and a change of frame of reference of deformation together have facilitated converting the moving base problem of earthquake shaking of buildings into a fixed base problem. Moving base problem can be solved by adopting appropriate Soil-Structure interaction model. Design codes recommended "design acceleration response spectrum", which is derived from the "acceleration response spectrum" of many individual ground motions. "design acceleration response spectrum" is different for each locations of the building site. It should be site specific. But, it would be tedious if designers are required to obtain this design spectrum by themselves for the design of individual buildings at a particular site. Soft soils are expected to shake more violently, and hence the "design acceleration response spectra" are different for these type of soil. "Design Acceleration Response Spectrum" recommended by codes is a spectrum recommended for use in the design of simple, regular and normal buildings. For the design of a special buildings, a "design acceleration response spectrum" should be developed for the specific building site. For site specific "design acceleration response spectrum", a huge data bank of soil properties need to be developed, which many countries are in a position to provide it to the structural engineers.

3.4.2   LIQUEFACTION POTENTIAL

The strength and stiffness of the soil may be reduced during earthquake-induced shaking, blasting, etc. It all depends on the properties of the particular soil. This phenomenon is known as *liquefaction*. It occurs in saturated, unconsolidated soils. Poorly drained fine-grained soils, such as sandy, silty, and gravelly soils, are the most susceptible to liquefaction. Granular soils are made up of a mix of soil and pore spaces. The water-filled pore spaces collapse during earthquake shaking, which leads to decreases in the overall volume of the soil. This process increases the water pressure between the individual soil grains, and the grains then move freely like a liquid, which substantially lowers the resistance to shear stress and causes the mass of soil to behave like a liquid. The soil starts to move in a downward direction due to the liquefaction.

In the liquefied state, soil deforms to a great extent, and a sudden loss of support takes place. Buildings may tilt and settle when liquefaction occurs. Liquefaction may also contribute to sand blowing, also known as *sand boils* or *sand volcanoes*. The density of the soil increases due to the collapse of its granular structure. The increased pressure squeezes the water out of the pore spaces between the soil grains and expels wet sand from the ground. The phenomenon of liquefaction is associated with a condition of zero effective stress, due to progressive increases in pore water pressure resulting from the tendency to densification of the sand structure subjected to cyclic loading. Catastrophic failures in past earthquakes have drawn serious attention to the fact that the liquefaction of sandy soils and sands with large amounts of non-plastic fine particles as a result of earthquake-induced ground shaking leads to a major threat to the safety of structures. Primary seismological factors that control liquefaction are the amplitude and frequency of the cyclic shear stress, as well as the duration of the shaking. The other site-specific factors that control the development of soil liquefaction are the grain size distribution of the soil mass, the relative density of the soil deposit, the depth and thickness of different soil strata, the depth of the ground water table, etc.

Pile caps rest on a number of piles. Due to liquefaction, skin frictional resistance is reduced, and as a result pile capacity is reduced suddenly during earthquake shaking. Soil provides lateral support to the piles. Due to liquefaction, the piles may undergo huge bending, perhaps leading to collapse. Therefore, the pile capacity should be obtained taking the liquefaction effect into account, and the design of Reinforced Concrete piles needs to be done accordingly.

A qualitative as well as quantitative understanding from laboratory investigation on the liquefaction process, have been reported. Pore water pressure generation and postliquefaction behavior in soils, has got special attention by the researchers. Many factors govern the liquefaction process are appreciated by them to some extent. However, the important parameters may be listed as follows: intensity of earthquake and its duration, location of ground water table, soil type, soil relative density, particle size gradation, particle shape, depositional environment of soil, soil drainage conditions, confining pressures, aging and cementation of the soil deposits, historical environment of the soil deposit and additional loads on these deposits.

Several approaches to evaluate the potential for liquefaction have been developed. The commonly employed are cyclic stress approach and cyclic strain approach to characterize the liquefaction resistance of soils both by laboratory and field tests. The cyclic stress approach to evaluate liquefaction potential characterizes both earthquake loading and the soil liquefaction resistance in terms of cyclic stresses. But, in the cyclic strain approach, earthquake loading and liquefaction resistance are characterized by cyclic strains. Cyclic tri-axial test, cyclic simple shear test and cyclic torsional shear test are the common laboratory tests. Further, standard penetration test (SPT), cone penetration test, shear wave velocity method, dilatometer test etc. are some of the in-situ tests to characterize the liquefaction resistance. Even though, cyclic stress and cyclic strain approaches are most widely used in the field of geotechnical earthquake engineering, some other approaches such as energy dissipation and effective stress based response analysis and probabilistic approaches have been also developed. IS 1893-Part I–2016 recommended different methods to identify liquefiable soil layers.

Numerical Example 1

Geo-technical data for a site situated in India (seismic zone IV as per IS 1893, Part 1, 2016, Seismic zone factor 0.24) are presented in Table 3.1. Liquefaction potential analysis has been done using **SPT value**.

From Table 3.1, it is observed that SPT value (N) of different layers of soil varies from 7 to 52. Density of soil at different layers are also presented. Percent fine, ground water depth, density of ground water etc. are also furnished.

TABLE 3.1
SPT value (N), density, percentage fine etc. of different layers of soil

Soil Data

Depth (m)	N	Density of soil (γ) kN/m³	Soil classification	Percentage fine	Ground water level (m)	Density of water (γ_w) kN/m3	Seismic zone factor (z)
1.5	7	17	Gravelly sand/	4	1.5	9.8	0.24
3	12	17.1	sandy gravel	5			
4.5	11	17.1		5			
6	25	17.4	Dense/very	7			
7.5	27	17.45	dense brownish	8			
9	29	17.5	grey silty fine/	7			
10.5	28	17.48	medium grained	7			
12	30	17.52	gravelly sand	8			
13.5	31	17.55		8			
15	35	17.6		10			
16.5	36	17.65		10			
18	37	17.66		10			
19.5	40	17.7	Gravels,	11			
21	42	17.7	surrounded and	11			
22.5	45	17.72	angular cobbles	12			
24	46	17.72	of quartzite	12			
25.5	46	17.72		12			
27	47	17.73		12			
28.5	48	17.73		12			
30	52	17.75		12			

Steps followed to estimate factor of safety (FoS) of different layers of soil against liquefaction potential

Step 1:

Let us consider, soil layer at 1.5 m level from EGL.

Computation of SPT blow count N_{60} for a hammer efficiency of 60%. If equipment used is of non-standard type, N_{60} is to be revised by applying correction factor (C_{60}).

Correction factors as per Table 12, Appendix F of IS 1893 (Part 1): 2016, considering standard hammer, standard sampler rod length 3 m and 150 mm bore diameter.

$C_{HT} = 1$

$C_{HW} = 1$

$C_{SS} = 1$

$C_{RL} = 0.75$

$C_{BD} = 1.05$

$C_{60} = C_{HT} \times C_{HW} \times C_{SS} \times C_{RL} \times C_{BD} = 1 \times 1 \times 1 \times 0.8 \times 1.05 = 0.79$

$N_{60} = C_{60} \times N = 0.7875 \times 20 = 5.53$

N_{60} value all the layers of soil are computed and shown in Table 3.2

Step 2:

Computed N_{60} is normalized according to effective overburden pressure of 100 kPa using overburden correction factor C_N.

Considering, at a depth of 1.5 m

Total stress (σ_v) = 1.5 x 17.7 = 25.5 kN/m^2

*Effective stress (σ_{v0}') = 25.5 – (0) x 9.8 = 25.5 kN/m^2

*Ground water level is 1.5m below EGL

$C_N = (P_a)^{1/2} \times (1 / \sigma_{v0}')^{1/2} \leq \mathbf{1.7}$

Standard average atmospheric pressure (P_a) = 101.325 kN/m^2

$C_N = (101.325)^{1/2} \times (1 / 30.27)^{1/2} = 1.99 > 1.7$, So $C_N = 1.7$

$(N_1)_{60} = C_N \times N_{60} = 9.4$

$(N_1)_{60}$ for all the layers of soil are furnished in Table 3.3

Step 3:

Computation of Cyclic resistance ratio $CRR_{7.5}$ for earthquake magnitude of 7.5

Considering at a depth of 1.5 m

Corrected $(N_1)_{60}$, i.e., $(N_1)_{60CS}$ considering fineness content (FC) in percentage

$(N_1)_{60CS} = \alpha + \beta (N_1)_{60}$

From Table 3.1, FC = 4%

From IS 1893 (Part 1), 2016, $\alpha = 0$, $\beta = 1$

$(N_1)_{60CS} = 0 + 1 \times 9.4 = 9.4$

$CRR_{7.5} = 1/(34 - (N_1)_{60CS}) + (N_1)_{60CS}/135 + 50 / [10*(N_1)_{60CS} + 45]^2 - 1/200$

$CRR_{7.5} = 0.11$

$(N_1)_{60CS}$ and $CRR_{7.5}$ for all the layers are furnished in Table 3.4

TABLE 3.2
Values of N_{60} of different layers of soil

As per Table 12 of IS 1893 Part 1, 2016

Depth (m)	N	C_{HT} (considering standard hammer)	C_{HW} (Coefficient for non standard hammer weight)	C_{SS} (Coefficient for non standard sampler)	length of hammer (m)	C_{RL} (Coefficient for short rod length)	C_{BD} (considering 150 mm bore diameter)	C_{60}	N_{60}
1.5	7	1	1	1	3	0.75	1.05	0.79	5.53
3	12	1	1	1	4.5	0.75	1.05	0.79	9.48
4.5	11	1	1	1	6	0.85	1.05	0.89	9.79
6	25	1	1	1	7.5	0.85	1.05	0.89	22.25
7.5	27	1	1	1	9	0.95	1.05	0.998	26.946
9	29	1	1	1	10.5	0.95	1.05	0.998	28.942
10.5	28	1	1	1	12	1	1.05	1.05	29.4
12	30	1	1	1	13.5	1	1.05	1.05	31.5
13.5	31	1	1	1	15	1	1.05	1.05	32.55
15	35	1	1	1	16.5	1	1.05	1.05	36.75
16.5	36	1	1	1	18	1	1.05	1.05	37.8
18	37	1	1	1	19.5	1	1.05	1.05	38.85
19.5	40	1	1	1	21	1	1.05	1.05	42
21	42	1	1	1	22.5	1	1.05	1.05	44.1
22.5	45	1	1	1	24	1	1.05	1.05	47.25
24	46	1	1	1	25.5	1	1.05	1.05	48.3
25.5	46	1	1	1	27	1	1.05	1.05	48.3
27	47	1	1	1	28.5	1	1.05	1.05	49.35
28.5	48	1	1	1	30	1	1.05	1.05	50.4
30	52	1	1	1	31.5	1	1.05	1.05	54.6

TABLE 3.3
Values of $(N_1)_{60}$ of different layers of soil

Depth (m)	Total Stress (σ_v) kN/m²	Effective stress (σ'_{v0}) kN/m²	C_N	Standard average atmospheric pressure (Pa) in kN/m²	$(N_1)_{60}$
1.5	25.5	25.5	1.7	101.325	9.4
3	51.15	38.41	1.7	101.325	16.12
4.5	76.8	49.36	1.7	101.325	16.64
6	102.9	60.76	1.7	101.325	37.83
7.5	129.08	72.24	1.7	101.325	45.81
9	155.33	83.79	1.7	101.325	49.2
10.5	181.55	95.31	1.7	101.325	49.98
12	207.83	106.89	1.7	101.325	53.55
13.5	234.15	118.51	1.7	101.325	55.34
15	260.55	130.21	1.7	101.325	62.48
16.5	287.03	141.99	1.7	101.325	64.26
18	313.52	153.78	1.7	101.325	66.05
19.5	340.07	165.63	1.7	101.325	71.4
21	366.62	177.48	1.7	101.325	74.97
22.5	393.2	189.36	1.7	101.325	80.33
24	419.78	201.24	1.7	101.325	82.11
25.5	446.36	213.12	1.7	101.325	82.11
27	472.95	225.01	1.7	101.325	83.9
28.5	499.55	236.91	1.7	101.325	85.68
30	526.17	248.83	1.7	101.325	92.82

Step 4:
Computation of Cyclic stress ratio (CSR), Cyclic resistance ratio (CRR), corrected $CRR_{7.5}$ for overburden stress level and high initial static shear stress and Factor of safety (FoS)
Considering at a depth of 1.5 m
Total stress $(\sigma v) = 1.5 \times 17 = 25.5 kN/m^2$
* Refer Step 6 of Annex F of IS 1893 (Part 1) 2016, if $C_N \leq 1.7$

Effective stress$(\sigma'_{v0}) = 25.5 - (0) \times 9.8 = 25.5 \ kN/m^2$
*Ground water level is 1.5 m below EGL
Stress reduction factor $(r_d) = 1 - 0.00765 \ z \ (0 < z < 9.15 \ m) = 0.989$
$CSR = 0.65*(a_{max}/g) * (\sigma_v/\sigma'_{v0}) * r_d = 0.15$
MSF = Magnitude scaling factor = 1 for earthquake magnitude 7.5
$K\sigma$ = Correction factor for high overburden stress = 1 for depth < 15 m
$K\alpha$ = Correction factor for static shear stress = 1 for flat ground
$CRR = MSF*K\sigma*K\alpha* CRR_{7.5} = 0.11$
$FoS = CRR/CSR = 0.73 < 1.0$
So the soil layer at a depth of 1.5 m, is not liquefiable
FS < 1.2, So, earthquake related permanent deformation is high.
The values CSR, CRR and FoS for all the layers of soil are furnished in Table 3.5

TABLE 3.4
Values of $CRR_{7.5}$ of different layers of soil

Depth (m)	$(N_1)_{60}$	Percentage fine	α	β	$(N1)_{60CS}$	$CRR_{7.5}$
1.5	9.4	4	0	1	9.4	0.11
3	16.12	5	0	1	16.12	0.17
4.5	16.64	5	0	1	16.64	0.18
6	37.83	7	0.12	1.01	38.3283	0.05
7.5	45.81	8	0.3	1.01	46.5681	0.26
9	49.2	7	0.12	1.01	49.812	0.3
10.5	49.98	7	0.12	1.01	50.5998	0.31
12	53.55	8	0.3	1.01	54.3855	0.35
13.5	55.34	8	0.3	1.01	56.1934	0.37
15	62.48	10	0.87	1.02	64.5996	0.44
16.5	64.26	10	0.87	1.02	66.4152	0.46
18	66.05	10	0.87	1.02	68.241	0.47
19.5	71.4	11	1.21	1.03	74.752	0.52
21	74.97	11	1.21	1.03	78.4291	0.55
22.5	80.33	12	1.55	1.03	84.2899	0.6
24	82.11	12	1.55	1.03	86.1233	0.61
25.5	82.11	12	1.55	1.03	86.1233	0.61
27	83.9	12	1.55	1.03	87.967	0.63
28.5	85.68	12	1.55	1.03	89.8004	0.64
30	92.82	12	1.55	1.03	97.1546	0.7

Numerical Example 2

Geo-technical data for a site situated in India (seismic zone III as per IS 1893, Part 1, 2016 (Seismic zone factor 0.16) are furnished in Table 3.6. Liquefaction potential analysis have been made using shear wave velocity value.

Step 1:

Computation of CSR (Cyclic stress ratio)
Considering at a depth 3 m
Total stress $(\sigma_v) = 3 \times 16.9 = 50.7$ kN/m²
*Effective stress $(\sigma'_{v0}) = 50.7 - (0) \times 9.8 = 50.7$ kN/m²
*Ground water level is 3.1m below EGL.
Stress reduction factor $(r_d) = 1 - 0.00765 z$ (0 < z < 9.15m) = 0.977
CSR = $0.65 \times (amax/g) \times (\sigma_v / \sigma'_{v0}) \times r_d = 0.102$
Computation for every layers have been done and furnished in Table 3.7

TABLE 3.5
Values of CSR, CRR and FoS

Depth (m)	$a_{max}/g = z$	$CSR = 0.65*(a_{max}/g)*(\sigma_v/\sigma'_{v0})*r_d$	Stress reduction factor (rd) = $1 - 0.00765h$ ($0 < h < 9.15$ m) or $1.174 - 0.0267h$ ($9.15 < h < 23$ m)	MSF	$K_\sigma = 1$ (if depth < 15 m) or $(\sigma'_{v0}/P_a)^{(f-1)}$, f = 0.7	K_α	$CRR = MSF*K_\sigma*K_\alpha*CRR_{7.5}$	FoS	Remarks
1.5	0.24	0.15	0.989	1	1	1	0.11	0.73	L
3	0.24	0.2	0.977	1	1	1	0.17	0.85	L
4.5	0.24	0.23	0.966	1	1	1	0.18	0.78	L
6	0.24	0.25	0.954	1	1	1	0.05	0.2	L
7.5	0.24	0.26	0.943	1	1	1	0.26	1	NL
9	0.24	0.27	0.931	1	1	1	0.3	1.11	NL
10.5	0.24	0.27	0.894	1	1	1	0.31	1.15	NL
12	0.24	0.26	0.854	1	1	1	0.35	1.35	NL
13.5	0.24	0.25	0.814	1	1	1	0.37	1.48	NL
15	0.24	0.24	0.774	1	1	1	0.44	1.83	NL
16.5	0.24	0.23	0.733	1	1.51	1	0.6946	3.02	NL
18	0.24	0.22	0.693	1	1.34	1	0.6298	2.86	NL
19.5	0.24	0.21	0.653	1	1.24	1	0.6448	3.07	NL
21	0.24	0.2	0.613	1	1.17	1	0.6435	3.22	NL
22.5	0.24	0.19	0.573	1	1.11	1	0.666	3.51	NL
24	0.24	0.17	0.533	1	1.06	1	0.6466	3.8	NL
25.5	0.24	0.16	0.493	1	1.02	1	0.6222	3.89	NL
27	0.24	0.15	0.453	1	0.98	1	0.6174	4.12	NL
28.5	0.24	0.14	0.413	1	0.95	1	0.608	4.34	NL
30	0.24	0.12	0.373	1	0.93	1	0.651	5.43	NL

Note: L = Liquefiable layer, NL = Non liquefiable layer

TABLE 3.6
Soil data showing shear wave values

Depth (m)	Description of strata	Average shear wave velocities V_s (m/sec)	Density of soil (γ) kN/m³	Atmospheric pressure Pa (kN/m2)	Percentage fine	Ground water level (m)	Density of water (γ_w) kN/m3	Seismic Zone factor (z)
3	Very soft soil	60	16.9	101.325	30	3.1	9.8	0.16
6	Very soft soil	80	17	101.325	32			
9	Very soft soil	88	17.1	101.325	33			
12	Very soft soil	95	17.2	101.325	34			
15	Stiff soil	170	19	101.325	37			
18	Stiff soil	172	19.2	101.325	37			
21	Stiff soil	183	19.5	101.325	38			
24	Stiff soil	202	19.8	101.325	40			
27	Stiff soil	210	20	101.325	41			
30	Stiff soil	215	20.2	101.325	42			

TABLE 3.7
Values of CSR of different layers of soil

Depth (m)	Total stress (σ_v) kN/m²	Effective stress (σ'_{v0}) kN/m²	Stress reduction factor $(r_d) = 1 - 0.00765h$ $(0 < h < 9.15$ m$)$ or $1.174 - 0.0267h$ $(9.15 < h < 23$ m$)$	CSR $= 0.65 \times (a_{max}/g)$ $\times (\sigma_v/\sigma'_{v0}) \times r_d$
3	50.7	50.7	0.977	0.102
6	101.7	73.28	0.954	0.138
9	153	95.18	0.931	0.163
12	204.6	117.38	0.908	0.173
15	261.6	144.98	0.885	0.175
18	319.2	173.18	0.693	0.174
21	377.7	202.28	0.613	0.172
24	437.1	232.28	0.533	0.136
27	788.1	553.88	0.453	0.091
30	912.9	649.28	0.373	0.078

Step 2:
Computation of Cyclic resistance ratio $(CRR_{7.5})$ for earth quake magnitude of 7.5
Considering at a depth 3 m
Shear wave velocity after overburden stress correction
$V_{S1} = (Pa / \sigma'_{v0})^{0.25} \times VS = 71.34$ m/sec
As per clause 6(c) per IS 1893 (Part 1), 2016
Limiting upper value of $V_{S1}*$ for liquefaction occurrence (varies from 200 m/sec
for FC = 35% and 215 m/sec for FC = 5% or less)
"a" and "b" are curve fitting parameters: a = 0.022 and b = 2.8
$CRR_{7.5} = a (V_{S1} / 100)^2 + b [(1/ (V_{S1}* - V_{S1}) - 1/V_{S1}*] = 0.12$
Computation for every layers have been done and furnished in Table 3.8

Step 3:
Computation of Cyclic resistance ratio CRR (by correcting $CRR_{7.5}$ for high over-burden stress level and high initial static shear stress) and Factor of safety (FoS)
Considering, soil layer at a depth of 3 m,
MSF = Magnitude scaling factor = 1 for earthquake magnitude 7.5
Kσ = Correction factor for high overburden stress = 1 for depth < 15 m
Kα = Correction factor for static shear stress = 1 for flat ground
$CRR = MSF * K\sigma * K\alpha * CRR_{7.5} = 0.12$
FoS = CRR/CSR = 1.18 > 1.0, soil is not liquefiable
FS > 1.2, So, earth quake related permanent deformation is small.
Computation for every layers have been done and furnished in Table 3.9

TABLE 3.8
Values of CRR$_{7.5}$ of different layers of soil

Depth (m)	Overburden stress corrected shear wave velocity V$_{S1}$ (m/sec)	Limiting upper value of V$_{S1}$* for liquefaction occurrence (m/sec)	Curve fitting parameters		
			a	b	CRR$_{7.5}$
3	71.34	200–215	0.22	2.8	0.12
6	86.75	200–215	0.22	2.8	0.18
9	89.39	200–215	0.22	2.8	0.19
12	91.57	200–215	0.22	2.8	0.2
15	155.44	200–215	0.22	2.8	0.58
18	150.43	200–215	0.22	2.8	0.54
21	153.95	200–215	0.22	2.8	0.57
24	164.16	200–215	0.22	2.8	0.66
27	137.34	200–215	0.22	2.8	0.45
30	135.13	200–215	0.22	2.8	0.43

TABLE 3.9
Values of CRR and FoS of different layers of soil

Depth (m)	MSF	K$_\sigma$ = 1 (if depth < 15 m) or (σ'_{v0} / P$_a$)$^{(f-1)}$, f = 0.7	K$_\alpha$	CRR	FoS	Remarks
2	1	1	1	0.12	**1.18**	*NL
4	1	1	1	0.18	**1.3**	NL
15	1	1	1	0.19	**1.17**	NL
18	1	0.85	1	0.17	**0.98**	*L
21	1	0.81	1	0.4698	**2.68**	NL
24	1	0.78	1	0.4212	**2.42**	NL
27	1	0.6	1	0.342	**1.99**	NL
30	1	0.57	1	0.3762	**2.77**	NL

*L = Liquefiable layer, NL = Non liquefiable layer

Numerical Example 3

Geo-technical data for a site in India (seismic zone IV as per IS 1893, Part 1, 2016, Seismic zone factor 0.24) is furnished in Table 3.10. Liquefaction potential analysis is done using CPT values etc.

TABLE 3.10
Soil testing data showing CPT values of different layers of soil

Depth (m)	Cone tip resistance q_c (kN/m²)	Sleeve friction f_s (kN/m2)	Density of soil (γ) kN/m³	Density of water (γ_w) kN/m3	Ground water level (m)
1	5600	74.1	18.42	9.8	2.4
2	9062	96.3	18.42		
3	9528	68.1	18.43		
4	9700	68.7	18.44		
5	7260	36.7	18		
6	7343	33.3	17.9		
7	7694	34.5	18		
8	7506	25.4	17.7		
9	7723	28.4	18.75		
10	7593	27.2	18.6		
11	7560	32.6	17.9		
12	7503	34.2	17.95		
13	7658	29.2	17.97		
14	7332	28.1	17.8		
15	7200	25.2	17.75		
16	8035	26	18.45		
17	8242	27.2	18.46		
18	8520	24.6	18.46		
19	8852	26.8	18.47		
20	8866	26.9	18.47		
21	8600	27.5	18.48		
22	8510	28.5	18.48		
23	8420	29.1	18.49		
24	8730	29.4	18.5		
25	8820	28.2	18.53		

Step 1:
Computation of normalized dimensionless cone penetration resistance q_{CIN}
Considering soil layer at a depth of 5 m
Total Stress (il laye 91.71 kN/m²)
Effective stress $(ye_{v0}) = 91.71 - (5 - 2.4) \times 9.8 = 66.23$ kN/m²
Standard average atmospheric pressure (Pa) = 101.325 kN/m²
*Correction factor for overburden pressure $C_Q = (Pa/\sigma'_{v0})^n = (101.325/66.23)^1 = 1.53$
*n = 1 for clay
Normalized dimensionless cone penetration resistance
$q_{CIN} = C_Q (qc/Pa) = 1.53 \times (7260/101.325) = 109.63$
Computation for every layers have been done and furnished in Table 3.11

TABLE 3.11
Values of normalized dimensionless cone penetration resistance of different layers of soil

Depth	Total Stress (σ_v) kN/m$_2$	Effective stress (σ'_{v0}) kN/m$_2$	Standard average atmospheric pressure (P_a) in kN/m^2	Correction factor for overburden pressure $C_{Q=}$ $(P_a/\sigma'_{v0})^n$ (n = 1 for clay)	Normalized dimensionless cone penetration resistance $q_{CIN=}$ $C_Q(q_c/P_a)$
1	18.42	18.42	101.325	5.5	303.97
2	36.84	36.84	101.325	2.75	245.95
3	55.27	49.39	101.325	2.05	192.77
4	73.71	58.03	101.325	1.75	167.53
5	91.71	66.23	101.325	1.53	109.63
6	109.61	74.33	101.325	1.36	98.56
7	127.61	82.53	101.325	1.23	93.4
8	145.31	90.43	101.325	1.12	82.97
9	164.06	99.38	101.325	1.02	77.74
10	182.66	108.18	101.325	0.94	70.44
11	200.56	116.28	101.325	0.87	64.91
12	218.51	124.43	101.325	0.81	59.98
13	236.48	132.6	101.325	0.76	57.44
14	254.28	140.6	101.325	0.72	52.1
15	272.03	148.55	101.325	0.68	48.32
16	290.48	157.2	101.325	0.64	50.75
17	308.94	165.86	101.325	0.61	49.62
18	327.4	174.52	101.325	0.58	48.77
19	345.87	183.19	101.325	0.55	48.05
20	364.34	191.86	101.325	0.53	46.38
21	382.82	200.54	101.325	0.51	43.29
22	401.3	209.22	101.325	0.48	40.31
23	419.79	217.91	101.325	0.46	38.23
24	438.29	226.61	101.325	0.45	38.77
25	456.82	235.34	101.325	0.43	37.43

Step 2: Values of Cyclic resistance ratio $CRR_{7.5}$ for earth quake magnitude of 7.5
Considering the soil layer at a depth of 5 m
$F = f_s \times 100 / (q_c - \sigma_v) = 36.7 \times 100/(7260 - 92) = 0.512$
$Q = (q_c - \sigma_v)/ P_a \times (P_a/\sigma'_{v0})^n = ((7260 - 91.71)/101.325) \times (101.325/66.23)^1 = 108.23$
$Ic = \sqrt{(3.47 - \log Q)^2 + (1.22 - \log F)^2} = 2.08$
Correction factor for grain characteristics Kc, Ic>1.64
$Kc = - 0.403 Ic^4 + 5.581 Ic^3 - 21.63 Ic^2 + 33.75 Ic - 17.88 = 1.42$
$(q_{CIN})_{CS} = K_c \times q_{CIN} = 1.42 \times 109.63 = 155.67$
$CRR_{7.5} = 0.833 \times ((q_{CIN})_{CS}/1000) + 0.05$, if $0 < (q_{CIN})_{CS} < 50$
or

TABLE 3.12
Cyclic resistance ratio ($CRR_{7.5}$) of different layers of soil

Depth (m)	$F = f_s \times 100/(q_c - \sigma_v)$	$Q = (q_c - \sigma_v/Pa) \times (Pa/\sigma'_{v_0})^n$	$Ic = \sqrt{(3.47 - logQ)^2 + (1.22 - logF)^2}$	Correction factor to account for grain characteristics Kc	$(q_{CIN})_{CS} = Kc \times q_{CIN}$	$CRR_{7.5} = 0.833 \times ((qCIN)CS/1000) + 0.05$, if $0 < (qCIN)CS < 50$ or $93 \times (qCIN)CS/1000)^3 + 0.08$, if $50 \le (qCIN)CS < 160$
1	0.762	303.02	1.48	1	303.97	2.69
2	1.067	244.98	1.61	1	245.95	1.46
3	0.719	191.79	1.81	1.11	213.97	0.99
4	0.714	165.88	1.85	1.15	192.66	0.75
5	0.512	108.23	2.08	1.42	155.67	0.43
6	0.46	97.31	2.15	1.55	152.77	0.41
7	0.456	91.68	2.17	1.6	149.44	0.39
8	0.345	81.4	2.29	1.92	159.3	0.46
9	0.376	76.06	2.29	1.92	149.26	0.39
10	0.367	68.5	2.33	2.05	144.4	0.36
11	0.443	63.29	2.29	1.92	124.63	0.26
12	0.469	58.54	2.3	1.95	116.96	0.23
13	0.393	55.97	2.37	2.19	125.79	0.27
14	0.397	50.34	2.4	2.31	120.35	0.24
15	0.364	46.64	2.45	2.53	122.25	0.25
16	0.336	49.27	2.46	2.57	130.43	0.29
17	0.343	47.83	2.46	2.57	127.52	0.27
18	0.3	46.94	2.5	2.77	135.09	0.31
19	0.315	46.43	2.49	2.72	130.7	0.29
20	0.316	44.31	2.51	2.82	130.79	0.29
21	0.335	40.98	2.51	2.82	122.08	0.25
22	0.351	38.76	2.52	2.87	115.69	0.22
23	0.364	36.71	2.53	2.92	111.63	0.21
24	0.355	36.59	2.53	2.92	113.21	0.21
25	0.337	35.54	2.56	3.09	115.66	0.22

$CRR_{7.5} = 93 \times (q_{CIN})_{CS}/1000)^3 + 0.08$, if $50 \le (q_{CIN})_{CS} < 160$
$CRR_{7.5} = 93 \times (155.67/1000)^3 + 0.08 = 0.43$
Computation for every layers have been done and furnished in Table 3.12

Step 3:
Computation of Cyclic resistance ratio (CRR) after correction for high overburden stress level and high initial static shear stress
Considering soil layer at a depth of depth = 5 m
MSF = Magnitude scaling factor = 1 for earthquake magnitude 7.5

TABLE 3.13
Cyclic resistance ratio (CRR) of different layers of soil

Depth (m)	MSF	K_σ	K_α	CRR
1	1	1	1	2.69
2	1	1	1	1.46
3	1	1	1	0.99
4	1	1	1	0.75
5	1	1	1	0.43
6	1	1	1	0.41
7	1	1	1	0.39
8	1	1	1	0.46
9	1	1	1	0.39
10	1	1	1	0.36
11	1	1	1	0.26
12	1	1	1	0.23
13	1	1	1	0.27
14	1	1	1	0.24
15	1	1	1	0.25
16	1	0.88	1	0.26
17	1	0.86	1	0.23
18	1	0.85	1	0.26
19	1	0.84	1	0.24
20	1	0.83	1	0.24
21	1	0.81	1	0.2
22	1	0.8	1	0.18
23	1	0.79	1	0.17
24	1	0.79	1	0.17
25	1	0.78	1	0.17

K_σ = Correction factor for high overburden stress = 1 for depth < 15 m
K_α = Correction factor for static shear stress = 1 for flat ground
CRR = MSF x K_σ x K_α x $CRR_{7.5}$ = 0.43
Computation for every layers have been done and furnished in Table 3.13

Step 4:
Computation of CSR (Cyclic stress ratio) and FoS (Factor of safety)
Considering soil layer at a depth of 5 m
Stress reduction factor (r_d) = 1 − 0.00765 z (0 < z < 9.15 m) = 0.962
CSR = 0.65 × (a_{max}/g) × (σ_v/σ'_{v0}) × r_d = 0.208
FoS = CRR/CSR = 0.43/ 0.208 = 2.07 > 1
Therefore, this soil layer is not liquefiable
FS > 1.2, So, earth quake related permanent deformation is small.
Computation for every layers have been done and furnished in Table 3.14

TABLE 3.14
Values of CSR and FoS of different layers of soil

Depth (m)	a_{max} / $g = z$	Stress reduction factor $(r_d) = 1 - 0.00765h$ $(0 < h < 9.15$ m) or $1.174 - 0.0267h$ $(9.15 < h < 23$ m)	CSR = 0.65 x (a_{max}/g) x (σ_v/σ'_{v0}) x r_d	FoS	Remarks
1	0.24	0.992	0.155	17.35	NL
2	0.24	0.985	0.154	9.48	NL
3	0.24	0.977	0.171	5.79	NL
4	0.24	0.969	0.192	3.91	NL
5	0.24	0.962	0.208	2.07	NL
6	0.24	0.954	0.219	1.87	NL
7	0.24	0.946	0.228	1.71	NL
8	0.24	0.939	0.235	1.96	NL
9	0.24	0.931	0.24	1.63	NL
10	0.24	0.907	0.239	1.51	NL
11	0.24	0.88	0.237	1.1	NL
12	0.24	0.854	0.234	0.98	L
13	0.24	0.827	0.23	1.17	NL
14	0.24	0.8	0.226	1.06	NL
15	0.24	0.774	0.221	1.13	NL
16	0.24	0.747	0.215	1.21	NL
17	0.24	0.72	0.209	1.1	NL
18	0.24	0.693	0.203	1.28	NL
19	0.24	0.667	0.196	1.22	NL
20	0.24	0.64	0.19	1.26	NL
21	0.24	0.613	0.183	1.09	NL
22	0.24	0.587	0.176	1.02	NL
23	0.24	0.56	0.168	1.01	NL
24	0.24	0.533	0.161	1.06	NL
25	0.24	0.507	0.154	1.1	NL

L = Liquefiable layer, NL = Non liquefiable layer

3.5 ESTIMATION OF BASE SHEAR

3.5.1 VARIOUS ASPECTS OF BASE SHEAR

Earthquake shaking is random and time variant. The basic concept of the design codes provides a design equivalent to the static lateral inertia force induced by earthquake shaking. Initially the seismic design base shear (V_B) is calculated, and then it is distributed along the height depending on the participation of mass at different levels. The seismic hazard at the building site is covered by the seismic zone factor (Z). The shaking of a building is a combined effect of the energy carried by the earthquake at different frequencies and the natural periods of the building. An importance factor (I) is introduced indirectly to reduce damage. The response reduction factor (R) is larger for ductile buildings and smaller for brittle ones. In view of the uncertainties involved in such parameters as the zone factor, seismic acceleration, etc., the upper limit of the imposed deformation demand on the building is not known as a deterministic upper bound value. Thus, the design is not termed an "earthquake-proof design". Instead, the earthquake demand is estimated only on the basis of the concept of the probability of exceedance, and the design allowing for earthquake effects is termed an "earthquake-resistant design", against the probable value of the demand.

The characteristics of seismic ground vibrations at any location depend upon the magnitude of the earthquake, its depth of focus, the distance from the epicenter, the characteristics of the path of the seismic waves and the soil strata at the building site. Ground motions are predominantly horizontal in direction at locations far from the epicenter. Near the epicenter, ground motions in the vertical direction are extremely important. Therefore, it is important to know the past earthquake history of the building site. The response of a building to the ground motions is a function of the nature of the soil, the type of foundations, the duration and characteristics of the ground motions, etc.

The design methodology needs to ensure that buildings possess at least a minimum strength to withstand minor earthquake events, of the type that occur frequently, without damage and should also be able to resist moderate earthquakes without significant damage, except for some nonstructural damage. The ultimate aim of earthquake-resistant design is that buildings should be able to withstand a major earthquake without collapsing. The actual forces may be much greater than the design forces as per the code, but the inelastic behavior of the building may be made in such a way that the ductile behavior of the building is assured, following the recommendations of codes strictly regarding the detailing of reinforcements. This will save the building from total collapse.

The response reduction factor (R) is, basically, dependent on the overstrength factor (R_s), the ductility factor (R_μ) and the redundancy factor (R_r). Studies have been carried out to evaluate the response reduction factor for different types of structures using *nonlinear static pushover analysis*. Different types of frames are considered by the researchers – e.g. Reinforced Concrete moment-resisting frames, braced frames, frames with a shear wall, etc. – and these factors are also reflected in the codes of practices. The structural ductility and overstrength capacity are the crucial constituents in defining the response reduction factor. The response reduction factor can be expressed

$$R = Rs * R\mu * Rr * R\xi$$

where

R_s = strength factor
R_μ = ductility factor
R_ξ = damping factor
R_r = redundancy factor

The additional strength beyond the design strength is generally termed *overstrength*. All properly designed buildings following code recommendations possess overstrength. The sequential compliance of critical regions, material overstrength, strain hardening and capacity reduction factors are the facts behind the overstrength factor (R_s). The advantage of overstrength is that it can be used in the design by reducing seismic shear, leading to more economical designs. The confinement of concrete, the strength contribution of nonstructural elements and special ductile detailing also contribute to overstrength.

$$R_s = V_u/V_d$$

where

    V_u is the maximum base shear
    V_d is the design base shear

Ductility factor (R_μ)

The ductility of a building is its capacity to support large inelastic deformations without significant loss of strength or stiffness. Buildings with high ductility can withstand large deformations and allow the building to move under high potential strength and absorb a large amount of energy. The amount of inelastic deformation encountered by the structural system subjected to a given ground motion or lateral loading is given by the displacement ductility ratio μ, and it is represented by the ratio of maximum absolute displacement to its yield displacement:

$$μ = Δu/Δy$$

Damping factor (R_ξ)

The damping factor (R_ξ) accounts for the impact of "added" adhesive damping and is, essentially, applicable for structures provided with additional energy-dissipating devices. If such devices do not exist, then the value for the damping factor is considered to be equal to 1.0, and it is eliminated from the explicit components of the response reduction factor used in force-based design procedures.

Redundancy factor (R_r)

The redundancy factor (R_r) is related to repetitions in a lateral-load-resisting system. Moment-resisting frames, shear walls, braced frames, etc. are adopted as lateral-load-resisting systems. Hence, the repetitions in lateral-load-resisting systems rely upon the structural system considered. A structural system with multiple lines of lateral-load-resisting framing systems comes under the category of redundant structural systems, because the frames are outlined and described to transfer the earthquake-induced inertia forces to the foundations. For ordinary moment-resisting frame (OMRF) (not detailed to improve ductility), response reduction factor is 3, for specially moment-resisting frames SMRF, specially detailed to ensure high ductility (SMRF) is 5. A Reinforced Concrete building not only should dissipate a considerable amount of additional energy for its ductile behavior, but also it should be able to control the deformations for its lateral stiffness, during earthquake shaking. Force based design procedure adopted by different seismic design codes to be based on static or dynamic analyses of elastic models of the buildings using elastic design spectra. The codes anticipate that building will undergo inelastic deformations under strong seismic events and therefore such inelastic behavior is usually incorporated into the design by dividing the elastic spectra by "Response reduction factor (R)", which reduces the spectrum from its original elastic demand level to a design level, where, structural ductility and overstrength capacity, are plying major role.

3.5.2   ESTIMATION OF BASE SHEAR AS PER IS 1893 (PART 1), 2016

3.5.2.1   Equivalent Static Method

Zone Factor

It is not possible to predict in advance that where and when earthquake will be occurred and in what magnitude with what type of ground motion character. IS 1893 has provided a seismic zoning map which indicates the value of peak ground acceleration as zone factor to estimate base shear. Depending on location of the building site, seismic zone can be identified and corresponding zone factor to be used as per table 2 /seismic zones map (annex of IS 1893-2016), to calculate base shear for the building

Seismic Weight

As per clause 7.4.1 of IS 1893 (Part 1), 2016, the seismic weight of each floor is the sum of the dead load of floor – i.e., the appropriate contribution of the weight of the columns, walls and any other permanent elements from the stories above and below, the floor finishes, the ceiling plaster, etc. – and the appropriate amount of the imposed/live load on the floor, as per code recommendations.

The appropriate amount of the imposed/live load to be considered in terms of the seismic weight = K xdesign imposed/live load, as per IS 875 (Part 2), 2015
   where

   K = reduction percentage of imposed/live load, as per table 10 of IS 1893 (Part 1), 2016. Table 10 specifies an imposed load/live load on a floor of up to and including 3.0 kN/m²; the percentage of the imposed load/live load to be considered is to be 25 percent of the design imposed load/live load, as per IS 875 (Part 2), 2015; if the imposed load/live load on the floor is more than 3.0 kN/m², the reduction percentage of the imposed load/live load to be considered is to be 50 percent of the design imposed load/live load, as per IS 875 (Part 2), 2015. The column and wall weight between consecutive floors is to be distributed and added to the seismic weight of a floor.

The seismic weight of each floor will act as a lumped weight at the center of mass of the floor.

Design horizontal seismic coefficient (A$_h$)

As per clause 6.4.2 of IS 1893 (Part 1), 2016:

$$A_h = (Z/2)(S_a/g)/(R/I)$$

where

Z = zone factor, as per table 3 of IS 1893 (Part 1), 2016

I = importance factor, depending upon the usage of building (it has to be decided whether the structure will be designed with a higher level of safety or not; the code provides the value of the importance factor: clause 6.4.2 and table 8 of IS 1893 (Part 1), 2016)

R = response reduction factor. It is the primary objective to design and detail a building structure which will have a low probability of collapse but with acceptable limit of damages of structural elements for a given level of ground shaking. As damage of the structural elements to a certain extent is permitted, so design seismic force should be less than that of expected maximum force. Code provides value of response reduction factor to get a design base shear force (clause 7.2.6 and table 9 of IS 1893 (Part 1), 2016)

S_a/g = design acceleration coefficient, specified in the code for different types of soil (normalized with the peak ground acceleration corresponding to the natural period of the structure: T_a; damping for Reinforced Concrete buildings may be taken to be 5 percent (clause 6.4.2 and figure 2 of IS 1893 (Part 1), 2016)

where

T_a = fundamental translational natural period (as per clause 7.6.2 of IS 1893 (Part 1), 2016, for different types of structure).

Design base shear

As per clause 7.6.1 of IS 1893 (Part 1), 2016:

$$V_B = A_h W$$

where

A_h = design horizontal seismic coefficient, as discussed above
W = seismic weight of the building, as calculated previously

Distribution of base shear along height of the building (at different floor levels)

As per clause 7.6.3a of IS 1893 (Part 1), 2016:

$$Q_i = V_b w_i h_i^2 \Big/ \sum_{j=1}^{n} w_j h_j^2$$

where

Q_i = design lateral force at ith floor level
w_i = seismic weight of ith floor level
h_i = height of ith floor measured from base (generally, from the top of the pile cap
to the ith floor level)
n = number of floors, including roof

Generally, a Reinforced Concrete slab may be considered as a rigid diaphragm in analysis of the building frame.

As per clause 7.6.3b of IS 1893 (Part 1), 2016, in buildings whose floors are capable of providing rigid horizontal diaphragm action in their own plane, the design story shear is to be distributed to the various vertical elements of the lateral-force-resisting system in proportion to the lateral stiffness of these vertical elements.

3.5.2.2   Response Spectrum Method

The response spectrum method of analysis is performed using a site-specific design spectrum or a design spectrum recommended in clause 6.4.6 of IS 1893 (Part 1), 2016.

Step 1 Identify the seismic zone, depending on the location of the building site, and consider the zone factor (Z), using table 2 and the seismic zones map (annex of IS 1893(Part 1),2016).

Step 2 Calculate the seismic weight of the building (W), as per clause 7.4.2 of IS 1893 (Part 1), 2016; the seismic weight of floors is to be considered as per clause 7.4.3 of IS 1893 (Part 1), 2016.

Step 3 Develop mass and stiffness matrices of the building, adopting the spring-lumped mass model concept.

Step 4 Compute the modal frequencies and corresponding mode shapes.

Step 5 Compute the modal mass (M_k) of the kth mode, using the following relationship, considering n number of modes as decided (only dominant modes)

Step 6 Compute the modal participation factors (P_k) of the kth mode, for n number of modes as decided (only dominant modes)

Step 7 Compute the design lateral load (Q_k) at each floor level for each mode (i.e., for ith floor mode k), using the following relationship:

$$Q_k = A_h(k)\, Q_{ik}\, P_k\, W_i$$

As per clause 7.8.4.5c of IS 1893 (Part 1), 2016

$A_h(k)$ = Design horizontal earthquake acceleration coefficient for mode k of oscillation
Q_{ik} = Design lateral force at floor i in mode k
P_k = Mode participation factor of mode k
W_i = Seismic weight of floor i

Step 8 Compute the story shear in each mode (V_{ik}) acting at a story level in mode k, as per clause 7.8.5d of IS 1893 (Part 1), 2016

where

$$V_k = \sum Q_{ik}$$

3.6 P-Δ ANALYSIS

The P-Δ effect is associated with displacements of the members at their ends. The P-Δ effect is important in assessing the overall structural behavior of the structures under a significant axial load. The P-Δ effect may contribute to a loss of lateral resistance, dynamic instability, effective lateral stiffness decreases, reducing strength capacity. To consider the P-Δ effect directly, the gravity load should be taken into account during nonlinear analysis. P-Δ effects increase the lateral flexibility of members under compression and increase the lateral stiffness of members under tension. However, the lateral stiffness of the overall building is not changed appreciably under wind loading, as the increased flexibility of columns on one side is counteracted by the smaller effect (or even tension stiffening) on the other side.

3.7 DUCTILITY ASSESSMENT

Pushover analysis is generally carried out as a static nonlinear analysis concept, where by a building is subjected to gravity loading and a monotonic displacement-controlled lateral load that continuously increases through elastic and inelastic behavior till a failure condition or a pre-fixed displacement is reached. The lateral load may represent the range of base shear induced by earthquake loading, and its configuration may be proportional to the distribution of mass along the building height, mode shapes, etc. Generally, a static pushover curve that plots a strength-based parameter against deflection is available after computer analysis. For example, performance may relate the strength level achieved in certain members to the lateral displacement at the tip of the building, or the bending moment may be plotted against the plastic rotation. It can be done in various way. The results provide insights into the ductile capacity of the structural system, and indicate the sequence of the failure mechanism, the load level and the deflection at which failure ultimately occurs. If an individual structural element of a building structure yields or fails, the dynamic forces on the building are shifted to other structural elements. A pushover analysis simulates this phenomenon by applying loads until the weak link in the structure is found and then revising the model to incorporate the necessary changes in the analysis. A second iteration indicates how the loads have been redistributed. The structure is "pushed" again until the second weak link is identified. This process continues until a yield pattern for the whole structure under seismic loading has been identified. Pushover analysis is commonly used to evaluate the seismic capacity of a building structure designed to fulfill the requirements regarding ductility, as per code recommendations. It is extremely useful for the performance-based design of buildings.

3.8 REINFORCED CONCRETE BUILDINGS WITH UNREINFORCED MASONRY INFILL WALLS

The spaces between the beams and columns of buildings are filled with unreinforced masonry infill. These infill elements play a part in the lateral response of buildings, and as a consequence alter the lateral stiffness of buildings. Hence, the natural periods and modes of oscillation of the building are affected in the presence of infill. In conventional design practice, the masses of the infill walls are taken into account, but their lateral stiffness is not. Modeling the infill wall in a building frame analysis is necessary to incorporate the additional lateral stiffness offered by infill walls. IS 1893(Part 1), 2016, suggests the infill be replaced by equivalent diagonal struts with a thickness equal to the thickness of the infill wall and a recommended width (a fraction of the diagonal length) and material properties. As a result, the lateral stiffness of the building increases when the infill walls are modeled as per this recommendation. As a result, the natural period of the building decreases. The extent of the stiffness enhancement and the change in the natural period due to the infill depends on the extent and spatial distribution of the infill. The change in the natural period is greater in shorter buildings compared to that in tall buildings. This implies that the seismic behavior of shorter buildings is affected more significantly than that of taller buildings when the stiffness enhancement due to infill is considered. Finally, the natural period has a direct effect on the estimation of the seismic base shear.

4 Structural Design of Reinforced Concrete Buildings

4.1 PREAMBLE

This chapter provides structural design and working drawings for two Reinforced Concrete buildings. Detailed step-by-step design calculations for a six-storied Reinforced Concrete building, considering dead, live, wind and seismic loads, are furnished. For this building, initially an equivalent static design approach as per IS code is followed, but the same building is also designed incorporating dynamic analysis under seismic conditions, and the results are compared. However, the dynamic effect of wind is not considered for this building, as it is not viewed as necessary in the IS code recommendations. The structural design of a 15-storied Reinforced Concrete building considering the dynamic effect due to both wind and seismic forces, as per IS code recommendations, is given. STAAD Pro CE software is used for this purpose. However, one can use any other reliable software package for this purpose. Design and detailing are made as per IS code provisions. **It is essential for the reader to have the relevant IS code of practices and handbooks for Reinforced Concrete design etc. as ready references while going through this chapter.**

4.1.1 STEPS FOR STRUCTURAL DESIGN OF REINFORCED CONCRETE FRAMED BUILDINGS

Step 1 Finalize the architectural plan, elevations, etc., incorporating the client's requirements as well as the structural stability, safety and economic aspects depending on wind, seismic, geotechnical and other relevant aspects as well.

Step 2 Finalize the column locations on the final architectural plan in consultation with the architect regarding aesthetics, functional aspects, etc., including the tentative shape and sizes of columns, beam, shear walls, etc.

Step 3 Prepare a structural plan showing the beams, columns, shear walls (if any), cutouts, drop slabs, etc.

Step 4 Calculate all the probable gravity loads acting on the slab panels, beams, etc. as per the code guidelines and the client requirements. Calculate the wind load, seismic load, temperature load, snow loads, etc. also as per the guidelines of the relevant codes.

Step 5 Develop a diagram showing the load dispersion from the slabs, either manually or using the output of the software employed.

Step 6 Calculate the wind and seismic loading at different levels/different nodes of the building frames on the longer and shorter faces of the building. Analyze manually, using the **substitute frame method**, as per IS 456, under gravity loading and the **cantilever method**, for horizontal loading such as wind and seismic loads. It may be noted here that all these methods are approximate ones. Rigorous analysis may be done using STAAD Pro CE software, or any other reliable software package.

Step 7 After getting the bending moments, shear forces, axial forces, displacements, etc., an exercise regarding load combinations has to be carried out as per the code guidelines so as to proceed to the final design of the Reinforced Concrete structural elements, in line with the guidelines of the relevant codes and handbooks.

Step 8 The foundations have to be designed against combinations of loads as per the code guidelines. The foundation type (shallow or deep foundations) may be finalized based on geotechnical reports and/or expert opinion.

Step 9 Prepare structural working drawings based on the design outputs and guidelines of the relevant codes and handbooks.

4.2   LIST OF RELEVANT IS CODES

1. IS 875 (Part 1), 1987 reaffirmed in 2018 provides unit weight of different building materials required for calculation of dead load
2. IS 875 (Part 2), 1987 reaffirmed in 2018 required for calculation of live load of different categories of building
3. IS 875 (Part 3), 2015 reaffirmed in 2020 required for calculation of wind load
4. IS 15498, 2015 for considering effects of cyclonic wind
5. IS 875 (Part 4), 1987 reaffirmed in 2018 for calculation of the snow load
6. IS 875 (Part 5), 1987 reaffirmed in 2018 for calculation of temperature, hydrostatic and soil pressure, other loads and load combinations
7. IS 1893 (Part 1), 2016 for calculation of the earthquake load on buildings
8. IS 4326, 2013 provides criteria for earthquake-resistant design and construction
9. IS 13920, 2016 provides guidelines for ductile design and detailing of Reinforced Concrete structures
10. IS 456, 2000 provides design guidelines for plain and reinforced concrete
11. SP16, 1999 a design aid for reinforced concrete design
12. Foundation design:
    a) IS 1904, 2006; IS 1080, 2002; IS 2950 (Part 1), 2008: guidelines for designing shallow foundations
    b) IS 2911 (Parts 1 to 4), 2010: guidelines for designing pile foundations

4.3   LOAD CALCULATION

4.3.1   DEAD LOAD

Self-weight of slab (kN/m^2) = thickness (m) x unit weight of Reinforced Concrete (kN/m^3)

Self-weight of beam rib (kN/m) = depth of beam rib (m) x width of beam (m) x unit weight of Reinforced Concrete (kN/m^3)

Self-weight of brick wall (kN/m) = clear height of wall*(m) x width of wall (m) x unit weight of brick wall

*clear height of wall = (floor to floor height – depth of floor beam rib)

Self-weight of floor finishing (kN/m^2) = thickness of floor finish × unit weight of floor material

Self-weight of concrete layer below tiles etc. (kN/m^2) = thickness of concrete layer × unit weight of laid concrete

Ceiling plaster weight (kN/m^2) = thickness of plaster x unit weight of plaster

The unit weight value may need to be taken from IS 875 (Part 1), 1987, or the handbook of the manufacturer.

4.4 DESIGN EXAMPLE OF A SIX-STORIED REINFORCED CONCRETE FRAMED RESIDENTIAL BUILDING

The relevant IS codes of practice will be required in order to follow the subsequent design calculations and working drawings.

Data

- A six-storied Reinforced Concrete framed residential building, as per the plan and sectional elevation shown respectively in Figures 4.1 and 4.2.
- The building is to be constructed at Kolkata, India.
- The grade of concrete used is M25 and the grade of steel used is Fe 500 TMT deformed bars.
- The typical floor height is 3,200 mm.
- The plinth height is 600 mm.
- The foundation depth is 1,500 mm.
- Walls other than brick partition walls are 250 mm thick, including plaster.

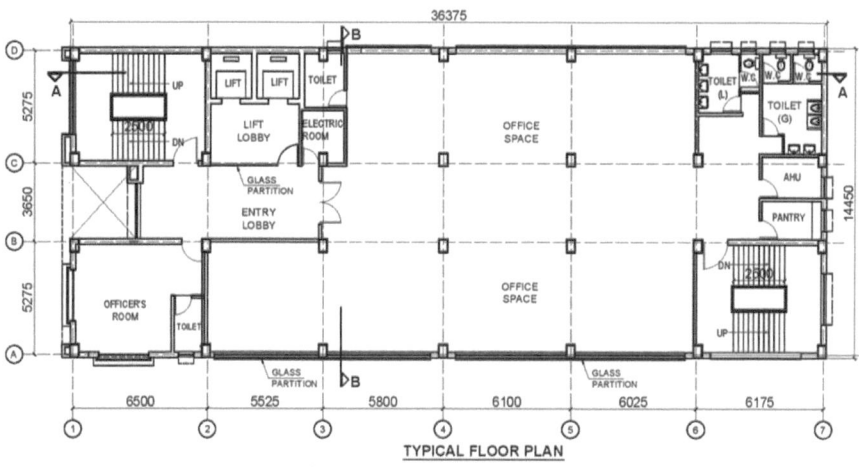

FIGURE 4.1 Plan of a six-storied building

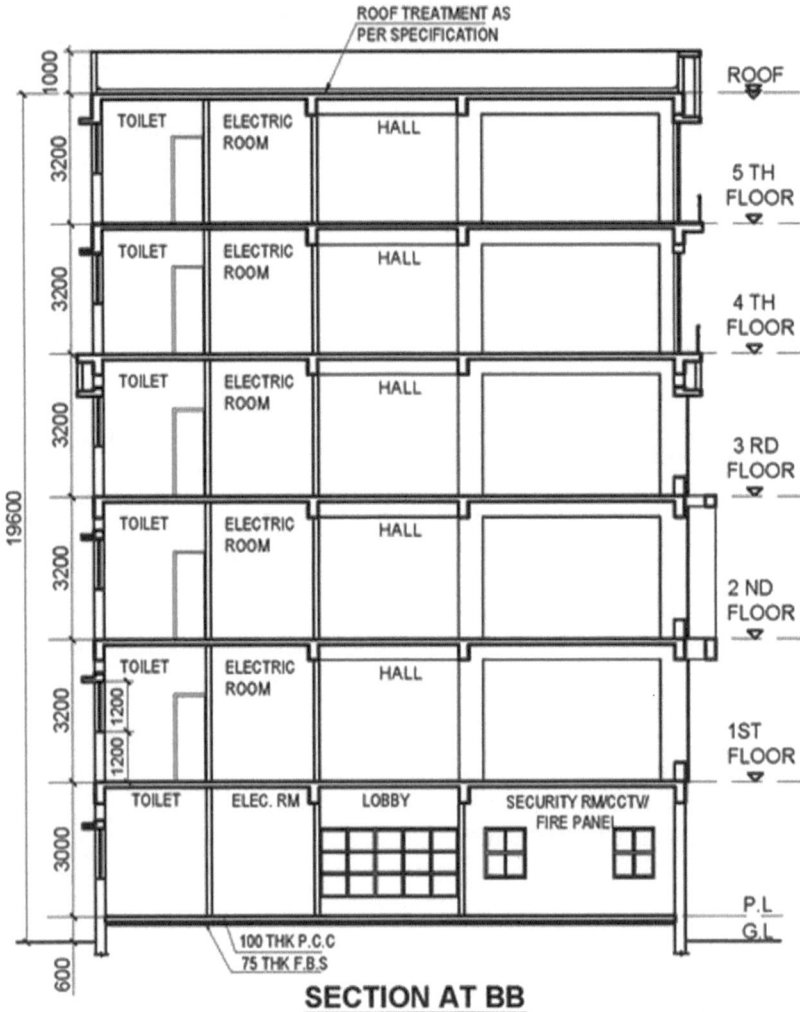

FIGURE 4.2 Sectional elevation of a six-storied building

- Brick partition walls are 125 mm, thick including plaster.
- The floor finish is 9.8 mm-thick vitrified tiles over 25 mm thick plain concrete.
- The ceiling plaster is 12 mm thick cement plaster.

4.4.1 CHOICE OF BEAM DEPTH

All parameters refer to Figure 4.1.

The maximum effective span between the columns in the shorter direction = 5,275 mm.

The maximum effective span between the columns in the longer direction = 6,500 mm.

As per the deflection criteria (clause 23.2.1 of IS 456, 2000):

Basic span to effective depth ratio = 26

As per figure 4 of IS 456, 2000:

$$f_s = 0.58(f_y)\{(\text{area of steel required})/(\text{area of steel provided})\}$$

Considering (area of steel required)/(area of steel provided) = 1

$$f_s = 0.58 \times 500 \times 1 = 290 \text{ N/mm}^2$$

Considering the percentage of tension steel $p_t = 0.24\%$, according to the criteria for the minimum percentage of tension steel as per IS 456, 2000 and IS 13920, 2016: the modification factor against tension reinforcement ($p_t = 0.24\%$) (from figure 4, IS 456) = 1.3.

Hence, the effective depth required to fulfill the deflection criteria as per clause 23 of IS 456, 2000, for beams in the shorter direction = 5,275/(26 x 1.3) = 156 mm.

The effective depth required to fulfill the deflection criteria as per clause 23 of IS 456, 2000, for beams in the longer direction = 6,500/(26 x 1.3) = 192 mm.

However, based on experience, the overall depth (D) for beams in the shorter direction may be considered, approximately, as span/10: i.e., = 5,275/10 = 528 mm ≈ 550 mm.

In other words, the effective depth (d) = 550 – 35 – (16/2) = 507 mm, allowing clear cover = 35 mm and diameter of reinforcement = 16 mm.

Similarly, the overall depth (D) for beams in the longer direction = 6,500/10 = 650 mm.

In other words, the effective depth (d) = 650 – 35– (16/2) = 607 mm.

- Note that the overall depth (D) ranges from **L/8 to L/15**, depending on dead, live, wind and seismic loads, the temperature load, the snow loads, the height of the building, the number of stories, the number of bays, the storey height, the plan ratio, irregularities, etc. The overall depth may be decided after consultation with experts and architectural considerations.

4.4.2 Choice of Slab Thickness

All parameters refer to Figure 4.4, for slab panels P1, P2, P3, P6, P7, P11, P12, P13, P14 and P15.

The maximum shorter span (l_x) of slab panels (l_x) = 3.25 m and the corresponding longer span (l_y) of slab panels = 5.275 m.

$$l_y/l_x = 5.275/3.25 = 1.62 < 2$$

Therefore, they are two-way slab panels.

With reference to clause 24.1 of IS 456, 2000, all slab panels are bounded with framed beams; therefore, they may be considered as continuous slab panels.

Then the shorter span (l_x)/effective depth (d) ratio = (40 x 0.8) = 32, allowing for HYSD/TMT bars/deformed bars.

Hence, the effective depth (d) = 3,250/32 = 102 mm.

Referring to clauses 26.4.2 and 26.4.3, and tables 16 and 16A, of IS 456, 2000, the recommended nominal cover of slab panels = 20 mm.

Therefore, the overall depth = 102 + 20 + 4 = 126 mm; let us say 140 mm, considering 8-mm-diameter reinforcing bars.

Similarly, for floor slab panels P8, P9 and P10, the maximum shorter span (l_x) = 3,650 mm and the requirement for overall depth = 140 mm.

Similarly, for floor slab panels P4 and P5, the maximum shorter span (l_x) = 5,275 mm and the requirement for overall depth = **190 mm.**

4.4.3   CALCULATION OF DEAD LOAD

All parameters refer to Figure 4.4, for slab panels marked P1 to P3 and P6 to P15.

For a Typical Floor Slab Panel

Self-weight of slab (140 mm thick) = (0.14) (25) =   3.5 kN/m²

Floor Finish

Self-weight of 25 mm-thick concrete below tiles = (0.025) (24) =   0.6 kN/m²
Self-weight of 9.8 mm-thick vitrified tiles =   0.185 kN/m²
Self-weight of 12 mm ceiling plaster = (0.012) (20.4) =   0.24 kN/m²
Light partition load directly on slab panels =   1.0 kN/m²*

≈ 5.55 kN/m²

  * In case the brick walls are placed directly on slab panels, an approximate equivalent distributed load has to be added as follows (though brick walls need to be considered as line load, and, accordingly, moment calculations need to be made). This is an approximate method. Rigorous plate theory may be applied in this regard. Calculations may be carried out to make it simpler.

$$*W_{bke} = \frac{1.25 \times Total\ weight\ of\ brick\ walls\ on\ a\ particular\ slab\ panel}{Area\ of\ the\ slab\ panel}$$

For Floor Slab Panels P4 and P5

Self-weight of slab (190 mm thick) = (0.19) (25) =   4.75 kN/m²

Floor Finish

Self-weight of 25 mm-thick concrete below tiles = (0.025) (24) =   0.6 kN/m²
Self-weight of 9.8 mm-thick vitrified tiles =   0.185 kN/m²
Self-weight of 12 mm ceiling plaster = (0.012) (20.4) =   0.24 kN/m²

Light partition load =   1.0 kN/m²

                                                                                          ≈ 6.8 kN/m²

For Roof Slab Panels P1 to P3 and P6 to P15

Self-weight of slab (140 mm thick) = (0.14) (25) =   3.5 kN/m²
Self-weight of 75 mm-thick waterproofing concrete = (0.075) (24) =   1.8 kN/m²
Self-weight of 12 mm ceiling plaster = (0.012) (20.4) =   0.24 kN/m²

                                                                                         ≈ 5.54 kN/m²

For Roof Slab Panels P4 and P5

Self-weight of slab (190 mm thick) = (0.19) (25) =   4.75 kN/m²
Self-weight of 75 mm-thick screed concrete = (0.075) (24) =   1.8 kN/m²
Self-weight of 12 mm ceiling plaster = (0.012) (20.4) =   0.24 kN/m²

                                                                                         ≈ 6.79 kN/m²

Dead Load of RCC Slab, including Floor Finish and Ceiling Plaster, in kN/m²

Self-weight of slab (kN/m²) = (0.1) (25) = 2.5 kN/m²
Self-weight of 25 mm-thick concrete layer below tiles etc. = (0.025) (24) =   0.6 kN/m²
Self-weight of 9.8 mm-thick vitrified tiles =   0.185 kN/m²
Self-weight of ceiling plaster 12 mm thick = (0.012) (20.4) =   0.24 kN/m²

----

                                                                                   w_d ≈ 3.5 kN/m²

Dead Load of Brick Walls, in kN/m

There is considered to be a perfect bond between unreinforced masonry work and the Reinforced Concrete frame elements (beams and columns). Therefore, in calculating the seismic weight, the weight of one-half of the walls above and below the beams has been taken into account.

Self-weight of 250 mm-thick wall = ((3.2 + 3.6)#/2 − 0.55) × 0.25 × 18.85 =
                                                13.43 kN/m
Self-weight of 125 mm-thick wall = ((3.2 + 3.6)/2 − 0.55) × 0.125 × 18.85 =
                                                 6.71 kN/m
The average of the ground floor height (3.6 m, including the plinth) and the next floor height (3.2 m) has been used.

Dead Load of Brick Walls/Beam Ribs, in kN/m, as Line Load

Self-weight of the rib of a beam = (0.25) (0.3) (25) = 1.875 kN/m
Self-weight of 250 mm-thick brick wall = (0.25) (3.2 − 0.3) (18.85) = 13.7 kN/m

4.4.4 LIVE/IMPOSED LOAD

The loads are as per IS 875 (Part 2), 1987 reaffirmed in 2018, table 1(i), for residential buildings.

Live load on all rooms, kitchen, toilet and bathroom = 2 kN/m²
Corridors, passages, staircases including fire escapes and store rooms, balconies = 3 kN/m²

As per IS 875 (Part 2), 1987 reaffirmed in 2018, Table 1(v), for business and office buildings.

Live load on a typical floor = 4.0 kN/m² (rooms without storage facility)
Live load on a corridor, staircase, etc. = 4.0 kN/m²
Live load on a typical roof slab panel as per IS 875 (Part 2), 1987 reaffirmed in 2018, table 2(1)(i) = 1.5 kN/m²
(taking into account an accessible roof)

4.4.5 APPROXIMATE AXIAL LOAD ON A PARTICULAR COLUMN

The *influence area method* is generally used at this stage of the calculation. It is an approximate approach. The "influence area" is defined as an area around the particular column on a "50% load sharing basis", and it is assumed that the entire load, including brickworks within the influence area, will be transferred to that particular column as shown in Figure 4.3.

Influence area = {(A/2) + (B/2)} x {(C/2) + (D/2)}

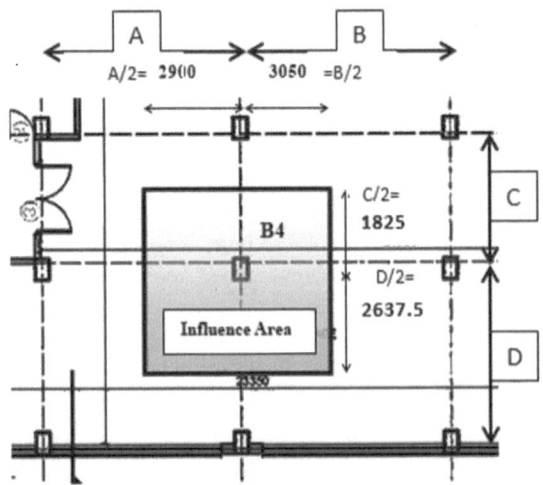

FIGURE 4.3   Influence area for the column marked B4

Approximate Load on Column Marked B4

A_f = Influence area of column marked B4 in a typical floor

$$= (2.9 + 3.05) (1.825 + 2.638) = 26.55 \text{ m}^2$$

Dead load transferred from floor slab = (influence area A_f) X (intensity of floor load w_d)

$$= * (5.55 + 6.8)/2 \times (26.55) = 164 \text{ kN}$$

* The thickness of few slab panels are 140 mm and the thickness of the remaining slab panels are 190 mm. For simplicity, an average load has been calculated.

At the ground floor column, the dead load is transferred from all five typical floor levels

$$= 5 \times 163.54 = 820 \text{ kN}$$

At the ground floor column, the wall load is transferred from five typical floor levels.

$$* = (10.42) (6.71) \times 5 = 350 \text{ kN}$$

* However, on grid 4 there is no partition wall in the plan, but for future provision partition wall loads have been taken into account.

The length of the wall/beam within the influence area = (2.9 + 3.05 + 1.825 + 2.64) = 10.42 m.

At the ground floor column, the load is transferred on account of the floor beams.

$$*= (10.42) [(0.55 + 0.65)/2 - 0.125] (0.25) (25) \times 5 = 155 \text{ kN}$$

The load transferred from the roof

$$= (26.55)(5.55 + 6.8)/2 = 164 \text{ kN}$$

The total axial load on the column marked B4 = 164 + 820 + 350 + 155 + 164 = 1,653 kN.

From experience, load may be increased by 25 percent to take into account the effect of moments at this stage of calculation.

Therefore, at this stage, we take the axial load on column B4 = 1.25 × 1,653 = 2,066 kN.

Applying the "limit state method", as per IS 456, 2000

$$\textbf{Factored load } (P_u) = 1.5 \times 2,066 = 3,099 \text{ kN}$$

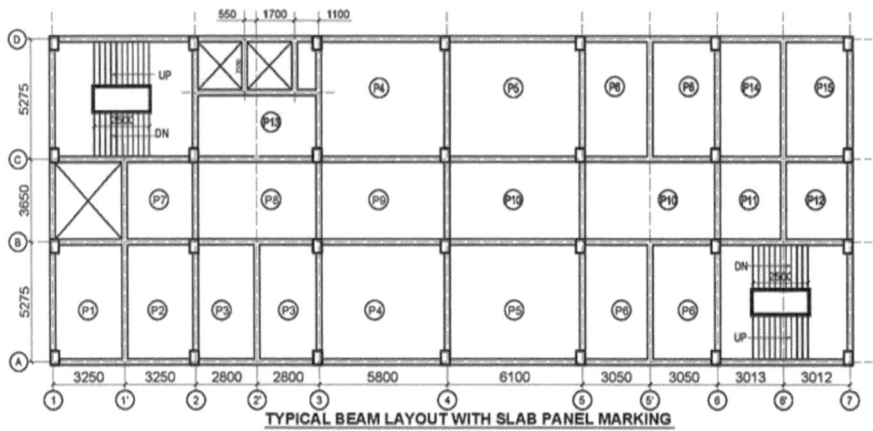

FIGURE 4.4   Marking of slab panels

Considering the percentage of steel = 1% and from chart 26 of SP-16,
Gross cross-sectional area of column B4 = 2,100 cm².

Let us provide a column section of 300 mm × 700 mm for column B4, for the time
being, till we get the exact design load and moments due to the dead load, live load,
wind load, seismic load, etc.; the section may need to be revised once the exact design
load and moments have been obtained.

Normally the longer side of a particular column is kept along the shorter side of
the building, to arrive at an economic design for that column. Keeping this concept
in mind, the longer side of the column – i.e., 700 mm – is kept along the shorter direction of the building, as shown in Figure 4.4. This kind of decision should be taken
by a structural engineer in consultation with the architect so as to avoid any conflict
between them.

4.4.6   DESIGN OF SLAB PANELS

Let us use clause 37.1.2 and Annex D of IS 456, 2000.

Shorter span moment coefficient:
   Positive coefficient = α_x
   Negative coefficient = $\alpha_x{}'$

Longer span moment coefficient:
   Positive coefficient = α_y
   Negative coefficient = $\alpha_y{}'$

The slab panels are marked according to the edge condition as per table 26 of IS
   456, 2000, presented in Table 4.1.

TABLE 4.1
Slab panels are marked according to edge condition (as per table 26 of IS 456, 2000)

Sl no.	Panel marked	Description	Case no.
1	P8, P9, P10, P11	Interior panel	4
2	P2, P3, P14	One short edge discontinuous	2
3	P4, P5, P7	One long edge discontinuous	3
4	P1, P6, P12, P13, P15	Two adjacent edges discontinuous	4

For Panels Marked P1 to P3 and P6 to P15

Dead load for a typical floor = 5.55 kN/m²
Live load for a typical floor = 4.00 kN/m²

$$= 9.55 \text{ kN/m}^2$$

For Panels Marked P4 and P5

Dead load for typical floor = 6.80 kN/m²
Live load for typical floor = 4.00 kN/m²

$$= 10.80 \text{ kN/m}^2$$

For Panels Marked P14 and P15

Load due to drop slab in the toilet = (0.125) (19) = 2.4 kN/m²
Dead load for typical floor   = 5.55 kN/m²
Live load for typical floor   = 4.00 kN/m²

$$= 11.95 \text{ kN/m}^2$$

The bending moment coefficients, according to Table 26 of IS 456, 2000, and the corresponding design bending moments for slab panels are presented in Table 4.2.
The design calculations are based on Tables 4.2, 4.3 and 4.3 (Continued).

• 8 mm – Fe 500 TMT steel reinforcement is used

A Few Important Aspects of the Design and Detailing of RCC Slabs

• Nominal cover as per Table 16A for a continuous slab with two hours fire resistance = 20 mm.
• Minimum reinforcement of a slab as per clause 26.5.2.1 = 0.12% (for HYSD/TMT bar), but as per clause 6.2.1 of IS 13920, 2016(for a flexural member) = 0.24%, considering the slab as a flexural member.
• Maximum spacing of reinforcements in tension as per clause 26.3.3 = 3d or 300 mm, whichever is less, where d = effective depth of slab
• Maximum diameter of reinforcing bar as per clause 26.5.2.2 should be less than D/8, where D = total depth of slab

TABLE 4.2
Ultimate design bending moment for slab panels (as per table 22 of IS 456, 2000)

Slab panel marked	l_y	l_x	l_y/l_x	(DL + LL)	Case no.	Bending moment coefficient				Positive bending moment in shorter span M_{ux}	Positive bending moment in longer span M_{uy}	Negative bending moment in shorter span M_{ux}'	Negative bending moment in longer span M_{uy}'
						α_x	α_y	α_x'	α_y'				
	m	m		(kN/m²)						kNm	kNm	kNm	kNm
P1	5.275	3.250	1.62	9.55	4	0.05484	0.035	0.07932	0.047	8.30	5.30	12.00	7.11
P2	5.275	3.250	1.62	9.55	2	0.035	0.028	0.06036	0.037	5.30	4.24	9.13	5.60
P3	5.275	2.800	1.88	9.55	2	0.05008	0.028	0.0666	0.037	5.62	3.14	7.48	4.16
P4	5.800	5.275	1.10	10.80	3	0.0325	0.028	0.0433	0.037	14.65	12.62	19.52	16.68
P5	6.100	5.275	1.16	10.80	3	0.0366	0.028	0.0488	0.037	16.50	12.62	22.00	16.68
P6	5.275	3	1.76	9.55	4	0.05924	0.035	0.08428	0.047	7.64	4.51	10.87	6.06
P7	3.65	3.25	1.12	9.55	3	0.069	0.028	0.0456	0.037	10.44	4.24	6.90	5.60
P8	5.6	3.65	1.53	9.55	1	0.04148	0.024	0.05384	0.032	7.92	4.58	10.28	6.11
P9	5.8	3.65	1.59	9.55	1	0.069	0.024	0.05524	0.032	13.17	4.58	10.54	6.11
P10	6.100	3.650	1.67	9.55	1	0.04372	0.024	0.05776	0.032	8.34	4.58	11.02	6.11
P11	3.650	3.013	1.21	9.55	2	0.0363	0.028	0.0484	0.037	4.72	3.64	6.29	4.81
P12	3.650	3.012	1.21	9.55	4	0.0395	0.035	0.0605	0.047	5.13	4.55	7.86	6.11
P13	5.600	2.640	2.00	9.55	4	0.065	0.035	0.091	0.047	6.49	3.49	9.09	4.69
P14	5.275	3.013	1.75	11.95	2	0.06684	0.028	0.064	0.037	10.88	4.56	10.41	6.02
P15	5.275	3.012	1.75	11.95	4	0.0554	0.035	0.084	0.047	9.01	5.69	13.66	7.64

TABLE 4.3
Design calculations for reinforcement for slab panels

Panel marked	Revised overall depth (D) mm	Revised effective depth (d) mm	M_{ux}/bd^2 N/mm²	M_{uy}/bd^2 N/mm²	M_{ux}'/bd^2 N/mm²	M_{uy}'/bd^2 N/mm²	Percentage of steel along shorter bottom p_{tx}	Percentage of steel along longer bottom p_{ty}	Percentage of steel along shorter top p_{tx}'	Percentage of steel along longer top p_{ty}'
P1	125.00	101	0.81	0.52	1.18	0.70	0.24	0.24	0.29	0.24
P2	125.00	101	0.52	0.42	0.90	0.55	0.24	0.24	0.24	0.24
P3	125.00	101	0.55	0.31	0.73	0.41	0.24	0.24	0.24	0.24
P4	190.00	166	0.53	0.46	0.71	0.61	0.24	0.24	0.24	0.24
P5	190.00	166	0.60	0.46	0.80	0.61	0.24	0.24	0.24	0.24
P6	125.00	101	0.75	0.44	1.07	0.59	0.24	0.24	0.26	0.24
P7	125.00	101	1.02	0.42	0.68	0.55	0.25	0.24	0.24	0.24
P8	140.00	116	0.59	0.34	0.76	0.45	0.24	0.24	0.24	0.24
P9	140.00	116	0.98	0.34	0.78	0.45	0.24	0.24	0.24	0.24
P10	140.00	116	0.62	0.34	0.82	0.45	0.24	0.24	0.24	0.24
P11	125.00	101	0.46	0.36	0.62	0.47	0.24	0.24	0.24	0.24
P12	125.00	101	0.50	0.45	0.77	0.60	0.24	0.24	0.24	0.24
P13	125.00	101	0.64	0.34	0.89	0.46	0.24	0.24	0.24	0.24
P14	125.00	101	1.07	0.45	1.02	0.59	0.26	0.24	0.25	0.24
P15	125.00	101	0.88	0.56	1.34	0.75	0.24	0.24	0.33	0.24

TABLE 4.3 (Continued)
Design calculations for reinforcement for slab panels

Panel marked	Area of steel				Spacing of reinforcement required				Bar dia.	Spacing of reinforcement provided			
	Shorter bottom A_{stx} mm²	Longer bottom A_{sty} mm²	Shorter top A'_{stx} mm²	Longer top A'_{sty} mm²	Shorter bottom S_x mm	Longer bottom S_y mm	shorter top S'_x mm	longer top S'_y mm		shorter bottom S_x mm	longer bottom S_y mm	shorter top S'_x mm	longer top S'_y mm
P1	242.4	242.4	292.9	242.4	206	206	170	206	8T	200	200	150	200
P2	242.4	242.4	242.4	242.4	206	206	206	206	8T	200	200	200	200
P3	242.4	242.4	242.4	242.4	206	206	206	206	8T	200	200	200	200
P4	398.4	398.4	398.4	398.4	125	125	125	125	8T	125	125	125	125
P5	398.4	398.4	398.4	398.4	125	125	125	125	8T	125	125	125	125
P6	242.4	242.4	262.6	242.4	206	206	190	206	8T	200	200	175	200
P7	252.5	242.4	242.4	242.4	198	206	206	206	8T	175	200	200	200
P8	278.4	278.4	278.4	278.4	179	179	179	179	8T	175	175	175	175
P9	278.4	278.4	278.4	278.4	179	179	179	179	8T	175	175	175	175
P10	278.4	278.4	278.4	278.4	179	179	179	179	8T	175	175	175	175
P11	242.4	242.4	242.4	242.4	206	206	206	206	8T	200	200	200	200
P12	242.4	242.4	242.4	242.4	206	206	206	206	8T	200	200	200	200
P13	242.4	242.4	242.4	242.4	206	206	206	206	8T	200	200	200	200
P14	262.6	242.4	252.5	242.4	190	206	198	206	8T	175	200	175	200
P15	242.4	242.4	333.3	242.4	206	206	150	206	8T	200	200	150	200

- As per clause D-1.4 of annex D, IS 456, 2000, the tension reinforcement provided at mid-span in the middle strip is to extend in the lower part of the slab to within 0.25L of a continuous edge or 0.15L of a discontinuous edge. L = clear span – i.e., the face-to-face distance between columns – and curtailment distances should be measured from the face of the column.
- As per clause D-1.5 of annex D, IS 456, 2000, over the continuous edges of a middle strip the tension reinforcement is to extend in the upper part of the slab a distance of 0.15L from the support, and at least 50 percent is to extend a distance of 0.3L.
- As per clause D-1.6 of annex D, IS 456, 2000, at a discontinuous edge, negative moments may arise depending on the degree of fixity at the edge of the slab panel but, in general, tension reinforcement equal to 50 percent of that provided at mid-span extending 0.1L into the span will be sufficient.

The above aspects are taken into account, and, accordingly, a rational detailing of slab reinforcements is shown in Figures 4.40 and 4.41.

4.4.7 Wind Load Analysis

4.4.7.1 Basic Wind Pressure
The detailed methodology has already been discussed in the previous chapter.
Place of construction: Kolkata, India.
Therefore, the basic wind speed is as per figure 1 of IS 875 (Part 3), 2015 = 50 m/s.

A Few Wind-Related Parameters
As per clause 6.3.1 and Table 1 of IS 875 (Part 3), 2015, the risk coefficient k_1 = 1.0 for this residential building, allowing for a return period of 50 years.

The site is located in a place where there are numerous closely spaced buildings around 10 m high with a few tall buildings. As per clause 6.3.2.1, it comes under terrain category 3.

It is to be noted that k_2 depends upon the height and the terrain category. Basic wind speed is calculated at 10 m height and terrain category 2. As per clause 6.3.2.2 and Table 2 of IS 875 (Part 3), 2015, the values of k_2 at different heights are given in Table 4.4.

As per clause 6.3.3.1 of IS 875 (Part 3), 2015, the topography factor k_3 = 1.0, allowing for an upwind slope < 3^0, considering the site is more or less leveled.

TABLE 4.4
Values of k_2 at different heights

height (m)	k_2
10	0.91
15	0.97
20	1.01
30	1.06

As Kolkata is not within a cyclonic region, k_4, therefore, is not applicable in this case.

Design Wind Velocity

As per clause 6.3 of IS 875 (Part 3) 2015:

$$V_z = (V_b) (k_1) (k_2) (k_3)$$

where

V_z = design wind speed at height z in m/s

Values of V_z at different heights are shown in Table 4.5.
As per clause 7.2 of IS 875 (Part 3), 2015, the wind pressure (p_z) at height z:

$$p_z = 0.6V_z^2$$

Wind pressure values at different heights are furnished in Table 4.6.
As per clause 7.2.1 of IS 875 (Part 3), 2015, the design wind pressure:

$$p_d = K_d \cdot K_a \cdot K_c p_z$$

but p_d shall not be less than $0.7p_z$
As per clause 7.2.1 of IS 875 (Part 3), 2015, the wind directionality factor:

$$K_d = 0.9$$

TABLE 4.5
Values of k_1, k_2, k_3, V_b and V_z at different heights

Height (m)	V_b (m/s)	k_1	k_2	k_3	V_z (m/s)
10	50	1	0.91	1	45.5
15	50	1	0.97	1	48.5
20	50	1	1.01	1	50.5
30	50	1	1.06	1	53.0

TABLE 4.6
Values of V_z and p_z at different heights

Height (m)	V_z (m/s)	p_z (N/m²)
10	45.5	1,242
15	48.5	1,411
20	50.5	1,530
30	53.0	1,685

TABLE 4.7
Design pressures at different heights

Height (m)	p_z (N/m²)	k_d	k_a	k_c	p_d	$0.7p_z$	p_d (N/m²)
10	1,242	0.9	0.8	0.9	805	869	869
15	1,411	0.9	0.8	0.9	914	945	945
20	1,530	0.9	0.8	0.9	991	1,011	1,011
30	1,685	0.9	0.8	0.9	1,092	1,180	1,180

Length of building = 36.125 m, width of building = 14.2 m, height of building = 19.6 m

The tributary area in both the longer and the shorter direction is greater than 100 m². As per clause 7.2.2 and table 4 of IS 875 (Part 3), 2015, the area averaging factor:

$$k_a = 0.8$$

As per clause 7.3.13 of IS 875 (Part 3), 2015, the combination factor:

$$k_c = 0.9$$

Values of the design wind pressure (p_d) at different heights are furnished in Table 4.7.

4.4.7.2 Wind Load as per "Drag Coefficient Approach"

The detailed methodology has already been discussed in the previous chapter.
Total wind load on the building:

$$F = C_f \, A_e \cdot p_d$$

where

C_f = drag coefficient,
A_e = frontal area against wind
p_d = design wind pressure

Effective Frontal Area for Each Joint/Node

A similar concept to the "Influence area concept", adopted earlier to get the approximate load transferred on a building (discussed earlier), is used to get the effective frontal area for each joint/node against the wind load.

Considering A1 joint/node on one of the faces of the building frame, the frontal area to be considered on that joint/node is shown in Figure 4.5.

Here, the center to center distance between A and B on the grid is 5.275 m. Therefore, one side of the frontal area is (5.275/2)m and the other side of the

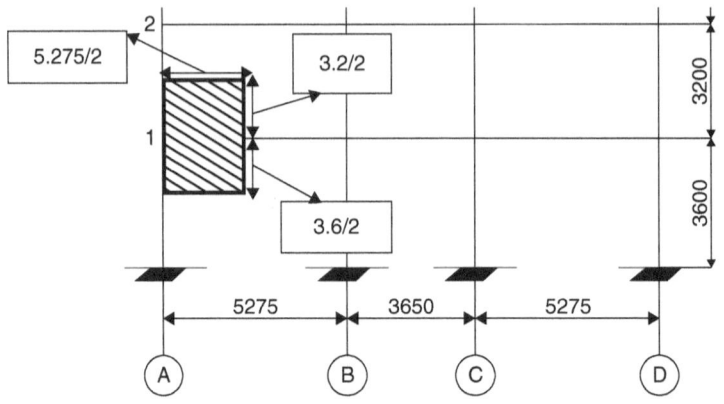

FIGURE 4.5   Frontal area for joint A1 against the wind load

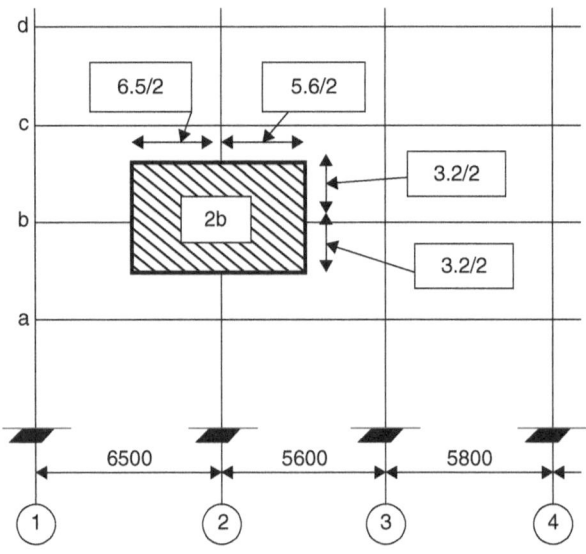

FIGURE 4.6   Influence area for joint 2b against the wind load

frontal area is along the height i.e., half the floor height above and below the joint/
node: [(3.2/2) + (3.6/2)]m.

Therefore, the frontal area against the wind for joint/node A1:

$$= [(5.275/2) \times \{(3.2/2) + (3.6/2)\}] = 5.98 \text{ m}^2$$

Let us consider joint/node 2b on the face of the frame, as shown in Figure 4.6.

Here, the center-to-center distance between 1 and 2 is 6.5 m, and for 2 and 3 the grid is 5.6 m. Therefore, one of the sides of the frontal area is (6.5/2 + 5.6/2)m and the other side of the frontal area is along the height – i.e., half the floor height above and below the joint/node: [(3.2/2) + (3.2/2)] m.

Therefore, the frontal area against the wind for joint/node 2b:

$$= [(6.5/2 + 5.6/2) \times \{(3.2/2) + (3.2/2)\}] = 12.91 \text{ m}^2$$

Details of the short and long frames are given in Figures 4.7 and 4.8.

The frontal areas of all the joints/nodes are calculated in a similar way, as furnished in Table 4.8.

Drag Coefficient (C_f)

The width of the building (a) = 14.2 m, the length of the building (b) = 36.125 m, the height of the building (h) = 19.6 m.

$$a/b = 0.4, \ h/b = 0.55$$

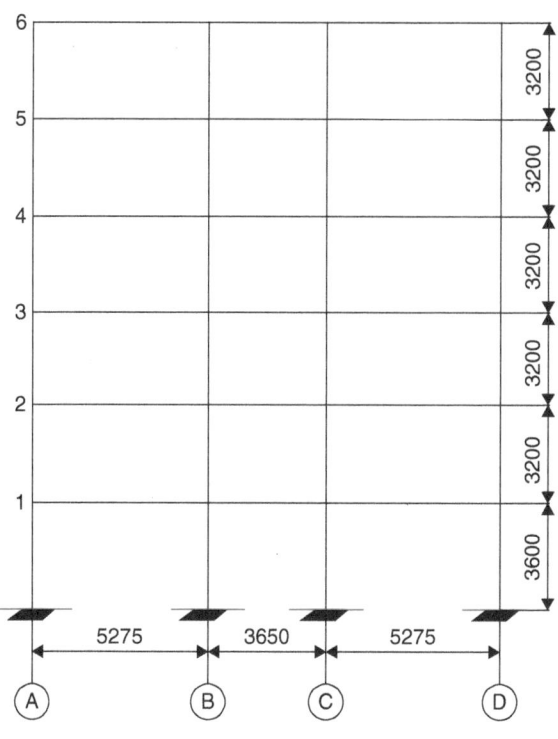

Short frame along 4 – 4

FIGURE 4.7   Building frame in the shorter direction (4–4)

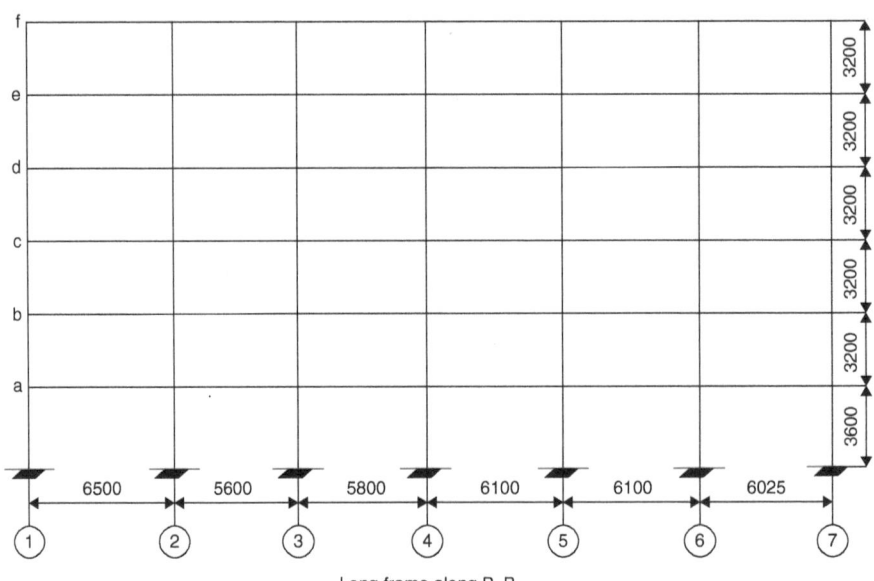

FIGURE 4.8   Building frame in the longer direction (B–B)

TABLE 4.8
Effective frontal area for each joint/node

Considering the shorter face of
the building

Joint/node marking	Width (m)	Height (m)	Frontal area – Ae (m²)
A1, D1	(5.275/2)	(3.6 + 3.2)/2	5.98
B1, C1	(5.275 + 3.65)/2	(3.6 + 3.2)/2	6.74
A6, D6	(5.275/2)	(3.2/2)	4.22
A2, A3, A4, A5, D2, D3, D4, D5	(5.275/2)	3.2	8.44
B2, B3, B4, B5, C2, C3, C4, C5	(5.275 + 3.65)/2	3.2	9.52
B6, C6	(5.275 + 3.65)/2	(3.2/2)	4.76
Considering the longer face of the building			
1a	(6.5/2)	(3.6 + 3.2)/2	7.37
2a	(6.5 + 5.6)/2	(3.6 + 3.2)/2	9.14
3a	(5.6 + 5.8)/2	(3.6 + 3.2)/2	8.61
4a	(5.8 + 6.1)/2	(3.6 + 3.2)/2	13.49
5a	6.1	(3.6 + 3.2)/2	13.83
6a	(6.1 + 6.025)/2	(3.6 + 3.2)/2	13.74
7a	(6.025)/2	(3.6 + 3.2)/2	6.83
1b, 1c, 1d, 1e	(6.5/2)	3.2	10.40
2b, 2c, 2d, 2e	(6.5 + 5.6)/2	3.2	12.91

(continued)

TABLE 4.8 (Continued)
Effective frontal area for each joint/node

Considering the shorter face of
the building

Joint/node marking	Width (m)	Height (m)	Frontal area – Ae (m²)
3b, 3c, 3d, 3e	(5.6 + 5.8)/2	3.2	12.16
4b, 4c, 4d, 4e	(5.8 + 6.1)/2	3.2	19.04
5b, 5c, 5d, 5e	6.1	3.2	19.52
6b, 6c, 6d, 6e	(6.1 + 6.025)/2	3.2	19.40
7b, 7c, 7d, 7e	(6.025)/2	3.2	9.64
1f	(6.5/2)	(3.2/2 + 1)	8.45
2f	(6.5 + 5.6)/2	(3.2/2 + 1)	10.49
3f	(5.6 + 5.8)/2	(3.2/2 + 1)	9.88
4f	(5.8 + 6.1)/2	(3.2/2 + 1)	15.47
5f	6.1	(3.2/2 + 1)	15.86
6f	(6.1 + 6.025)/2	(3.2/2 + 1)	15.76
7f	(6.025)/2	(3.2/2 + 1)	7.83

TABLE 4.9
Wind load on nodes/joints (on the shorter face of the building)

Joint marking	Height (m)	Frontal area (A_e) (m²)	Drag coefficient (C_f)	Design wind pressure (p_d) (kN/m²)	Wind load at the joint/node (kN)
A1, D1	3.6	5.98	1.2	0.869	6.24
B1, C1	3.6	6.74	1.2	0.869	7.03
A2, A3, D2, D3	6.8 & 10.0	8.44	1.2	0.869	8.8
B2, B3, C2, C3	6.8 & 10.0	9.52	1.2	0.869	9.93
A4, D4	13.2	8.44	1.2	0.945	9.57
B4, C4	13.2	9.52	1.2	0.945	10.8
A5, D5	16.4	8.44	1.2	1.011	10.24
B5, C5	16.4	9.52	1.2	1.011	11.55
B6, C6	19.6	4.76	1.2	1.071	6.12
A6, D6	19.6	4.22	1.2	1.071	5.42

As per clause 7.4.2.1 and figure 4 of IS 875 (Part 3), 2015:

$$C_f \approx 1.2$$

The wind loads on joints/nodes are calculated and furnished in Tables 4.9 and 4.10.

TABLE 4.10
Wind load on nodes/joints (on the longer face of the building)

Joint marking	Height (m)	Frontal area (A_e) (m^2)	Drag coefficient (C_f)	Design wind pressure (p_d) (kN/m^2)	Wind load at the joint/node (kN)
1a	3.6	7.37	1.2	0.869	7.69
2a	3.6	9.14	1.2	0.869	9.53
3a	3.6	8.61	1.2	0.869	8.98
4a	3.6	13.49	1.2	0.869	14.07
5a	3.6	13.83	1.2	0.869	14.42
6a	3.6	13.74	1.2	0.869	14.33
7a	3.6	6.83	1.2	0.869	7.12
1b, 1c	6.8 & 10.0	10.40	1.2	0.869	10.85
2b, 2c	6.8 & 10.0	12.91	1.2	0.869	13.46
3b, 3c	6.8 & 10.0	12.16	1.2	0.869	12.68
4b, 4c	6.8 & 10.0	19.04	1.2	0.869	19.85
5b, 5c	6.8 & 10.0	19.52	1.2	0.869	20.36
6b, 6c	6.8 & 10.0	19.40	1.2	0.869	20.23
7b, 7c	6.8 & 10.0	9.64	1.2	0.945	10.93
1d	13.2	10.40	1.2	0.945	11.79
2d	13.2	12.91	1.2	0.945	14.64
3d	13.2	12.16	1.2	0.945	13.79
4d	13.2	19.04	1.2	0.945	21.59
5d	13.2	19.52	1.2	0.945	22.14
6d	13.2	19.40	1.2	0.945	22.00
7d	13.2	9.64	1.2	0.945	10.93
1e	16.4	10.40	1.2	1.011	12.62
2e	16.4	12.91	1.2	1.011	15.66
3e	16.4	12.16	1.2	1.011	14.75
4e	16.4	19.04	1.2	1.011	23.10
5e	16.4	19.52	1.2	1.011	23.68
6e	16.4	19.40	1.2	1.011	23.54
7e	16.4	9.64	1.2	1.011	11.70
1f	19.6	8.45	1.2	1.071	10.86
2f	19.6	10.49	1.2	1.071	13.48
3f	19.6	9.88	1.2	1.071	12.70
4f	19.6	15.47	1.2	1.071	19.88
5f	19.6	15.86	1.2	1.071	20.38
6f	19.6	15.76	1.2	1.071	20.25
7f	19.6	7.83	1.2	1.071	10.06

4.4.7.3 Wind Load as per the "Pressure Coefficient Approach"

The detailed methodology has already been discussed in the previous chapter.

Considering individual structural elements of the building, such as walls: as per clause 7.3.1 of IS 875 (Part 3), 2015:

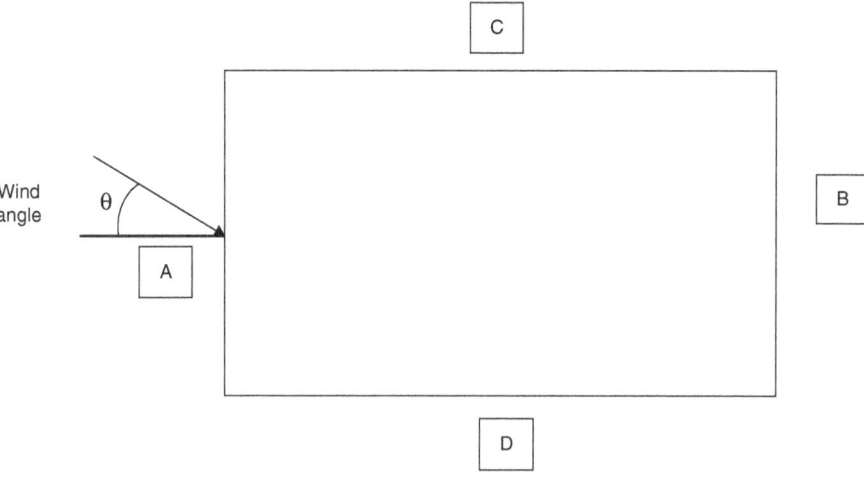

FIGURE 4.9 Plan showing the wind angle and external wall faces

Note: A, B, C and D indicate different faces of the building.

$$F = (C_{pe}\text{-}C_{pi}). \, A. \, p_d$$

where

F = wind force acting at a joint/node
A = frontal area against wind
C_{pe} = external wind pressure coefficient
C_{pi} = internal wind pressure coefficient
p_d = design wind pressure

External Wind Pressure Coefficient (C_{pe})

The width of the building (w) = 14.2 m, the length of the building (l) = 36.125 m, the height (h) of the building = 19.6 m.

$$l/w = 2.5, \, h/w = 1.4$$

As per Table 5 of IS 875 (Part 3), 2015, the external wind pressure coefficients for different faces of the building are shown as per Figure 4.9 and are furnished in Table 4.11.

For a building with a medium amount of openings – i.e., 5 to 20 percent of the wall area– the internal pressure coefficient C_{pi} = ±0.5, as per clause 7.3.2.2.

The net wind pressure coefficient for the different surfaces of the building are furnished in Table 4.12.

We know that

$$F = (C_{pe} - C_{pi}). \, A. \, p_d$$

TABLE 4.11
External wind pressure coefficients

C_{pe} for surfaces (when wind angle is 0°)

A	B	C	D
0.7	−0.3	−0.7	−0.7

C_{pe} for surfaces (when wind angle is 90°)

A	B	C	D
−0.5	−0.5	0.7	−0.1

TABLE 4.12
Net wind pressure coefficients for different surfaces

	C_{pe} for surfaces (when wind angle is 0°)			
	A	B	C	D
	+0.7	−0.3	−0.7	−0.7
	C_{pi} for medium openings			
Case 1	−0.5	−0.5	−0.5	−0.5
Case 2	+0.5	+0.5	+0.5	+0.5
$(C_{pe}-C_{pi})$ for Case1	+1.2	+0.2	−0.2	−0.2
$(C_{pe}-C_{pi})$ for Case2	+0.2	−0.8	−1.2	−1.2
	C_{pe} for surfaces (when wind angle is 90°)			
	A	B	C	D
	−0.5	−0.5	+0.7	−0.1
	C_{pi} for medium openings			
Case 1	−0.5	−0.5	−0.5	−0.5
Case 2	+0.5	+0.5	+0.5	+0.5
$(C_{pe}-C_{pi})$ for Case 1	0	0	+1.2	+0.4
$(C_{pe}-C_{pi})$ for Case 2	−1.0	−1.0	+0.2	−0.6

Therefore, the wind pressure taking into account the pressure coefficients given in Table 4.12 and the basic pressure given in Table 4.6 are calculated and provided in Tables 4.13 and 4.14.

The wind load, as per both the "drag coefficient method" and the "pressure coefficient method", has been considered and the input given to computer analysis using STAAD Pro CE. However, manual frame analysis using the "cantilever method" has been done with the load data obtained from the "drag coefficient method". Wind loads on short frame 4–4 are shown in Figure 4.10 and on the longer frame in Figure 4.11.

TABLE 4.13
Wind pressure on wall when θ = 0°

Height (m)	p_d (N/m²)	Wind angle θ = 0° Wind pressure on different surfaces (kN/m²)			
		A	B	C	D
Case 1					
10 m	869	+1.0400	+0.1738	–0.1738	–0.1738
15 m	945	+1.1300	+0.1890	–0.1890	–0.1890
20 m	1,011	+1.21	+0.2022	–0.2022	–0.2022
Case 2					
10 m	869	+0.1738	–0.6952	–1.0428	–1.0428
15 m	945	+0.189	–0.7560	–1.1340	–1.1340
20 m	1,011	+0.2022	–0.8088	–1.2132	–1.2132

TABLE 4.14
Wind pressure on wall when θ = 90°

Height (m)	p_d (N/m²)	Wind angle θ = 90° Wind force on different surfaces (kN/m²)			
		A	B	C	D
Case 1					
10	869	0	0	+1.0428	+0.3476
15	945	0	0	+1.1340	+0.3780
20	1,011	0	0	+1.2132	+0.4044
Case 2					
10	869	–0.869	–0.869	+0.1738	–0.5214
15	945	–0.945	–0.945	+0.1890	–0.5670
20	1,011	–1.011	–1.011	+0.2022	–0.6066

4.4.8 SEISMIC LOAD ANALYSIS

The detailed methodology of seismic load analysis has already been discussed in Chapter 3. Seismic analysis uses the "equivalent static method".

Calculation of Design Horizontal Seismic Coefficient (A_h)

As per clause 6.4.2 of IS 1893 (Part 1), 2016:

$$A_h = (Z/2)(S_a/g)/(R/I)$$

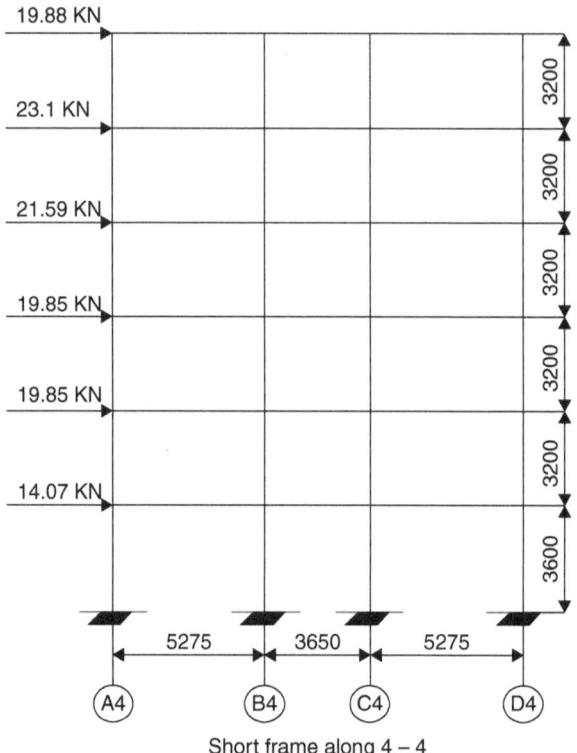

Short frame along 4 – 4

FIGURE 4.10   Wind loads on frame 4–4 (along the shorter direction of the building)

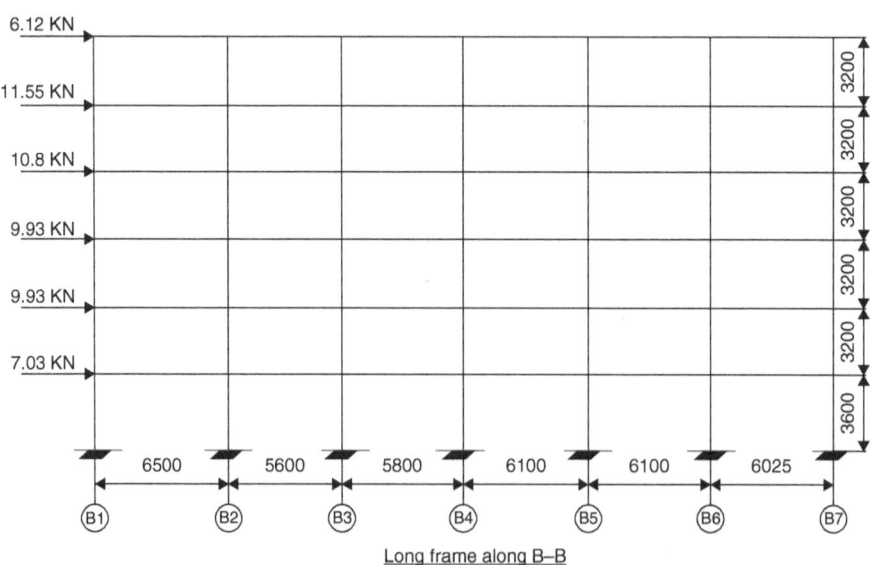

Long frame along B–B

FIGURE 4.11   Wind loads on frame B–B (along the longer direction of the building)

where

Z = zone factor (as per table 3 of IS 1893 (Part 1), 2016)

I = importance factor, depending upon the usage of the building (as per clause 6.4.2 and table 8 of IS 1893 (Part 1), 2016)

R = response reduction factor (as per clause 2.6 and table 9 of IS 1893 (Part 1), 2016)

S_a/g = design acceleration coefficient (as per clause 6.4.2 and figure 2 of IS 1893 (Part 1), 2016)

T_a = fundamental translational natural period (as per clause 7.6.2 of IS 1893 (Part 1), 2016)

As per clause 6.4.1 of IS 1893 (Part 1), 2016, and figure 1, Kolkata is in **Zone III**. As per clause 6.4.2 of IS 1893 (Part 1), 2016, the design horizontal seismic coefficient:

$$A_h = (Z/2)\ (S_a/g)/(R/I)$$

As per figure 1 and table 3 of IS 1893 (Part 1), 2016, the seismic zone factor:

$$Z = 0.16$$

As per clause 7.2.3 and table 8 of IS 1893 (Part 1), 2016, the importance factor for a normal commercial building:

$$I = 1.2$$

As per clause no 7.2.6 and Table 9 of IS 1893 (Part 1), 2016, the response reduction factor for a special moment-resisting frame (SMRF)

$$R = 5$$

As per clause 7.6.2(c) of IS 1893 (Part 1), 2016, the fundamental translational natural period:

$$T_a = 0.09h/\sqrt{d}$$

h = height of the building = 21.1 m (allowing 1.5 m from ground level to the top of the pile cap)

d = base dimension of the building at the plinth level along the direction of earthquake shaking

d in X direction = 36.125 m, d in Y direction = 14.2 m

Therefore, T_{ax} = 0.29 sec and T_{ay} = 0.47 sec

From figure 2A of IS 1893 (Part 1), 2016, allowance is to be made for foundations in medium soil and 5 percent damping for the Reinforced Concrete structure.

In this case, Sa/g = 2.5 for both directions, and therefore:

$$A_h = (0.16/2) \times (2.5)/(5/1.2) = 0.048$$

Seismic Weight

As per clause 7.4.1 of IS 1893 (Part 1), 2016, the seismic weight of each floor is the sum of the dead load of the floor – i.e., the appropriate contribution of the weight of the columns, walls and any other permanent elements from the stories above and below, the floor finishes, the ceiling plaster, etc. – and the appropriate amount of the imposed/live load on the floor as per the code recommendation.

Appropriate amount of imposed/live load to be considered on account of seismic weight:

$$= K \times \text{design imposed/live load (as per IS 875 (Part 2))}$$

where

K = reduction percentage of imposed/live load, as per Table 10 of IS 1893 (Part 1), 2016

Calculation of dead weights

With reference to Figure 4.4, the calculation of the weights at different levels is furnished in Tables 4.15, 4.16, 4.17, 4.18 and 4.19.

TABLE 4.15
Weight of roof slab

Slab panel	Number of panels	Lx = shorter side (m)	Ly = longer side (m)	Weight intensity of the slab panel (kN/m²)	Weight of the slab panel (kN)
Panel P1	1	3.250	5.275	5.54	94.98
Panel P2	1	3.250	5.275	5.54	94.98
Panel P3	2	2.800	5.275	5.54	163.65
Panel P4	2	5.275	5.800	6.80	416.09
Panel P5	2	5.275	6.100	6.80	437.61
Panel P6	4	3.050	5.275	5.54	356.53
Panel P7	1	3.250	3.650	5.54	65.72
Panel P8	1	3.650	5.600	5.54	113.24
Panel P9	1	3.650	5.800	5.54	117.28
Panel P10	2	3.650	6.100	5.54	246.70
Panel P11	1	3.013	3.650	5.54	60.93
Panel P12	1	3.012	3.650	5.54	60.91
Panel P13	1	3.275	5.600	5.54	101.60
Panel P14	1	3.013	5.275	5.54	88.05
Panel P15	1	3.012	5.275	5.54	88.02
Lift room	1	5.275	5.525	6.80	198.18
Stair room	2	5.275	6.500	6.80	466.31
Total (W_1)					**3,170.78**

TABLE 4.16
Weight of parapet walls, stair walls, water tank, etc. from roof level

Item	Number	Length (m)	Height (m)	Width (m)	Unit weight (kN/m³)	Weight (kN)
1 m-high parapet 250 thick		100.65	1.00	0.25	18.85	474.31
Stair wall		23.55	3.20	0.25	18.85	355.13
Lift wall		21.75	3.20	0.25	18.85	327.99
Beam rib		258.1	0.41	0.25	25.00	661.38
Column	28	2.60	0.30	0.70	25.00	382.20
PVC overhead tank (150kL capacity)						150.00
Total (W₂)						**2,351.01**

TABLE 4.17
Weight of floor slabs

Slab panel	Number of panels	Lx = shorter side (m)	Ly = longer side (m)	Weight intensity of the slab panel (kN/m²)	Weight of the slab panel (kN)
Panel P1	1	3.250	5.275	5.55	95.15
Panel P2	1	3.250	5.275	5.55	95.15
Panel P3	2	2.800	5.275	5.55	163.95
Panel P4	2	5.275	5.800	6.80	416.09
Panel P5	2	5.275	6.100	6.80	437.61
Panel P6	4	3.050	5.275	5.55	357.17
Panel P7	1	3.250	3.650	5.55	65.84
Panel P8	1	3.650	5.600	5.55	113.44
Panel P9	1	3.650	5.800	5.55	117.49
Panel P10	2	3.650	6.100	5.55	247.14
Panel P11	1	3.013	3.650	5.55	61.04
Panel P12	1	3.012	3.650	5.55	61.02
Panel P13	1	3.275	5.600	5.55	101.79
Panel P14	1	3.013	5.275	5.55	88.21
Panel P15	1	3.012	5.275	5.55	88.18
Total (W₃)					**2,509.27**

TABLE 4.18
Weight on account of walls, columns, etc. at a typical floor level

Item	Number	Length (m)	Height (m)	Width (m)	Unit weight (kN/m³)	Weight (kN)
External wall 250 mm thick		100.650	2.65	0.250	18.85	1256.93
Internal wall 125 mm thick		76.275	2.65	0.125	18.85	476.27
Internal wall 250 mm thick		36.875	2.65	0.250	18.85	460.50
Beam rib		258.100	0.41	0.250	25.00	661.38
Column	28	3.200	0.30	0.700	25.00	470.40
Total (W_4)						**3,325.48**

TABLE 4.19
Weight of walls, beams and column elements

Item	Number	Length (m)	Height (m)	Width (m)	Unit weight (kN/m³)	Weight (kN)
External wall 250 mm thick		60.85	1.60	0.25	18.85	458.81
Tie beam		258.10	0.25	0.45	25.00	725.91
Column	28	1.60	0.30	0.70	25.00	235.20
Total (W_5)						**1,419.92**

Total dead load of the building = W1 + W2 + W3 + W4 + W5

$$= 3,170.78 + 2,351 + 5(2,509.27 + 3,325.48) + 1,419.92 = \textbf{36,115 kN}$$

Live Load/Imposed Load

As per Table 10 and clause 7.3.2 of IS 1893 (Part 1), 2016, no live load/imposed load is considered at roof level and a 50 percent reduction is made.

Live load/imposed load from a typical floor = 0.5 x [(36.125 x 14.2) x 4] = 1,026 kN

Total live load/imposed load on account of seismic weight of the total building:

$$5 \times 1,026 = \textbf{5,130 kN}$$

Seismic weight of the total building = 36,115 + 5,130 = **41,245 kN**

Calculation of Design Base Shear

As per clause 7.6.1 of IS 1893 (Part 1), 2016, design base shear (V_B)

$$= A_h W = 0.048 \times 41,245 = \textbf{1,979.8 kN}$$

Distribution of base shear along the height of the building at different floor levels, as per clause 7.6.3A of IS 1893 (Part 1), 2016:

$$Q_i = V_b \left\{ w_i h_i^2 \Big/ \sum_{J=1}^{n} w_j h_j^2 \right\}$$

TABLE 4.20
Distribution of lateral forces (inertial force) and shear forces at different floor levels

Floor	w_i (kN)	h_i (m)	$w_i h_i^2$	V_B (kN)	Q_i	V_i (kN)
Roof	5,412	21.1	2,409,477.0	1,979.8	626.09	626.09
5th	6,817	17.9	2,184,235.0	1,979.8	567.56	1,193.65
4th	6,817	14.7	1,473,086.0	1,979.8	382.77	1,576.42
3rd	6,817	11.5	901,548.3	1,979.8	234.26	1,810.68
2nd	6,817	8.3	469,623.1	1,979.8	122.03	1,932.71
1st	6,817	5.1	177,310.2	1,979.8	46.07	1,978.78
Plinth	1,748	1.5	3,933.0	1,979.8	1.02	1,979.80
Total	41,245		$\sum w_j h_j^2 = 7,619,212.0$		$\sum Qi = 1,979.80$	

TABLE 4.21
Distribution of lateral forces (inertial force)

Floor	Q_i (kN)	Number of frames in X direction	Number of frames in Y direction	Q_i in X direction per frame (kN)	Q_i in Y direction per frame (kN)
Roof	626.09	7	4	89.44	156.52
5th	567.56	7	4	81.08	141.89
4th	382.77	7	4	54.68	95.69
3rd	234.26	7	4	33.47	58.57
2nd	122.03	7	4	17.43	30.51
1st	46.07	7	4	6.58	11.52
Plinth	1.02	7	4	0.15	0.26

where

Q_i = design lateral force at ith floor level
w_i = seismic weight of ith floor level
h_i = height of ith floor measured from base (generally, from the top of the pile cap to the ith floor level)
n = number of floors, including roof

The seismic weight of each floor will act as a lumped weight at the center of mass of the floor. Generally, a reinforced concrete slab may be considered as a rigid diaphragm during computer analysis of the building frame. (Later on, dynamic analysis is also done using STAAD Pro CE software.)

As per clause 7.6.3B of IS 1893 (Part 1), 2016, in buildings whose floors are capable of providing a rigid horizontal diaphragm action in their own plane, the design story shear is to be distributed to the various vertical elements of the lateral-force-resisting system in proportion to the lateral stiffness of these vertical elements. The design lateral seismic force in each floor is calculated in Tables 4.20 and 4.21. The seismic force per floor for short and long frames is shown in Figures 4.12 and 4.13.

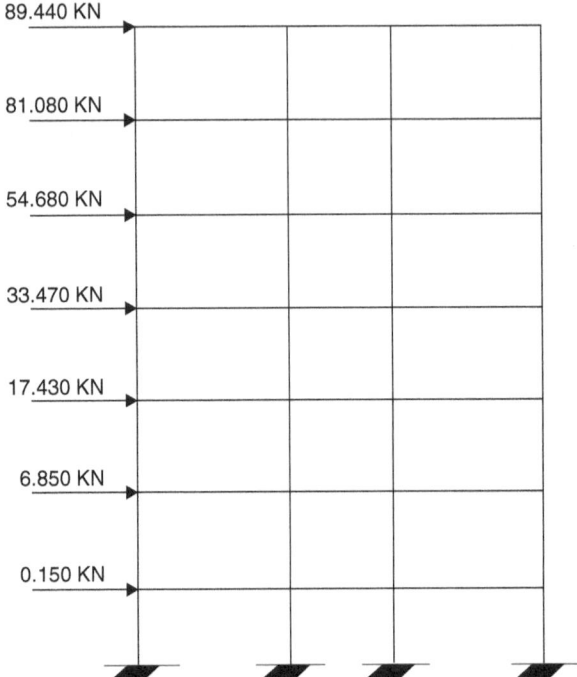

FIGURE 4.12   Frame in the shorter direction showing the seismic load in each joint

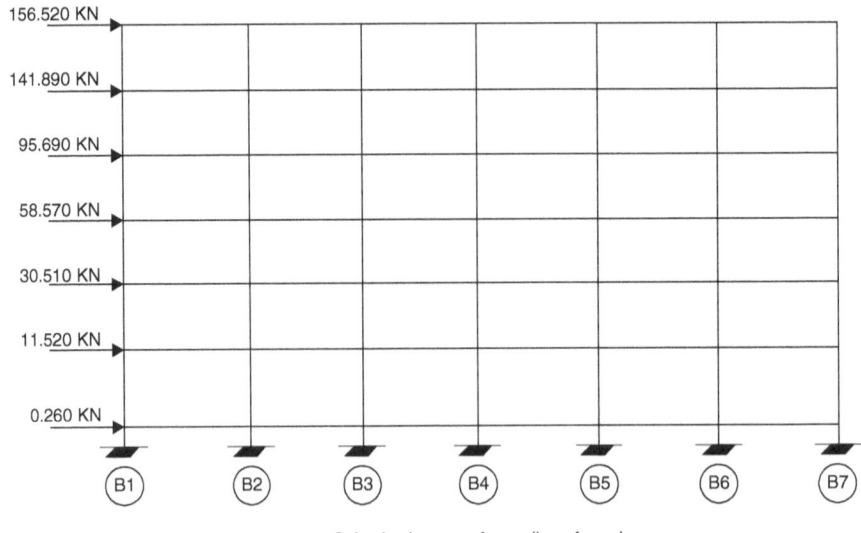

Seismic shear per frame (long frame)

FIGURE 4.13   Frame in the longer direction showing the seismic load in each joint

4.4.9 Substitute Frame Analysis under Dead and Live Loads

Load dispersion as per the rule of IS 456 is followed and shown in Figure 4.14.

For the shorter frame 4–4, and as per Figure 4.15 and the dispersion of the load as shown in Figure 4.14:

Column size = 300 mm x 700 mm and beam size = 550 mm x 250 mm

as obtained tentatively beforehand.

The stiffness and distribution factor calculations of the frame refer to Figure 4.15, and are provided in Tables 4.22 and 4.23.

The load dispersion on frame 4–4 is shown in a partly plan view in Figure 4.14, and the loads transferred are shown in Figure 4.16.

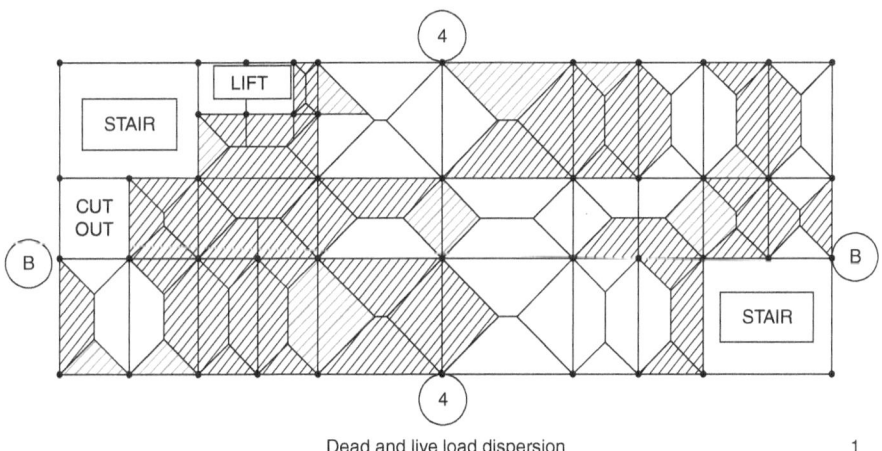

Dead and live load dispersion   1

FIGURE 4.14   Typical floor load dispersion for dead and live loads

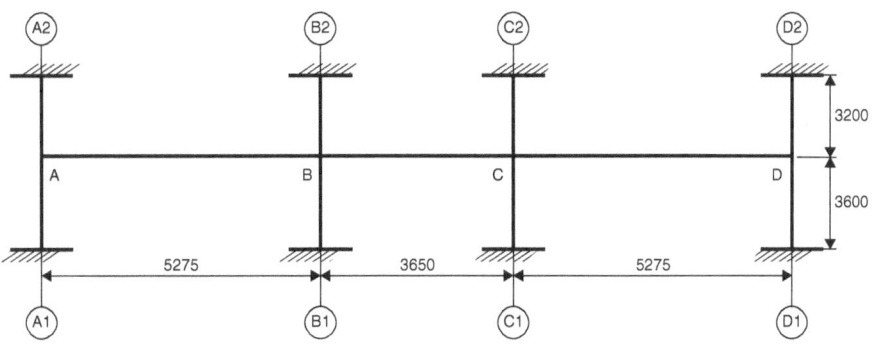

FIGURE 4.15   Substitute frame A1–B1–C1–D1 at first floor level in the shorter direction (4–4)

TABLE 4.22
Stiffness of members of the shorter frame (see Figure 4.15)

Member	Length (mm)	Moment of inertia of beam (I_B) (mm⁴)	Moment of inertia of column (I_C) (mm⁴)	Stiffness (K) = 4EI/L
AB, CD	5,275	I_B = (1/12) bd³	I_C = (1/12) bd³	$4EI_B/5{,}275$
BC	3,650	= (1/12) (250) x	= (1/12) (700) x	$4EI_B/3{,}650$
AA1, BB1, CC1, DD1	3,600	(550)³	(300)³	$4EI_C/3{,}600$
AA2, BB2, CC2, DD2	3,200	= 3,466,145,833	= 1,575,000,000	$4EI_C/3{,}200$

TABLE 4.23
Distribution factor (DF) at joints of the frame (see Figure 4.15)

Joint	Member	K	$\sum K$	DF=$K_{ij}/\sum K$
A	AB	2,628,357.03E	10,065,857.03E	0.261
	AA1	3,500,000E		0.348
	AA2	3,937,500E		0.391
B	BA	2,628,357.03E	13,864,373.01E	0.190
	BC	3,798,515.982E		0.274
	BB1	3,500,000E		0.252
	BB2	3,937,500E		0.284
C	CB	3,798,515.982E	13,864,373.01E	0.274
	C	2,628,357.03E		0.190
	CC1	3,500,000E		0.252
	CC2	3,937,500E		0.284
D	DC	2,628,357.03E	10,065,857.03E	0.261
	DD1	3,500,000E		0.348
	DD2	3,937,500E		0.391

The uniformly distributed load (UDL)

= 6.71 kN/m (125 mm-thick brick wall) +
3.43 kN/m (self-weight of the beam rib)

For AB and CD (190 mm-thick slabs) the load intensity (w) = (5.55 + 1.25) = 6.8 kN/m² , and for BC (a 140 mm-thick slab) the load intensity (w) = 5.55 kN/m² (refer to the load calculation).

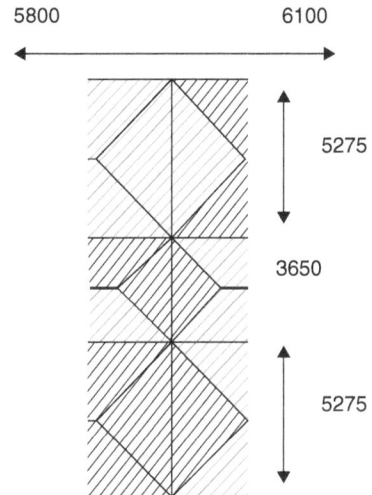

FIGURE 4.16 Dispersion of floor loads on frame 4–4

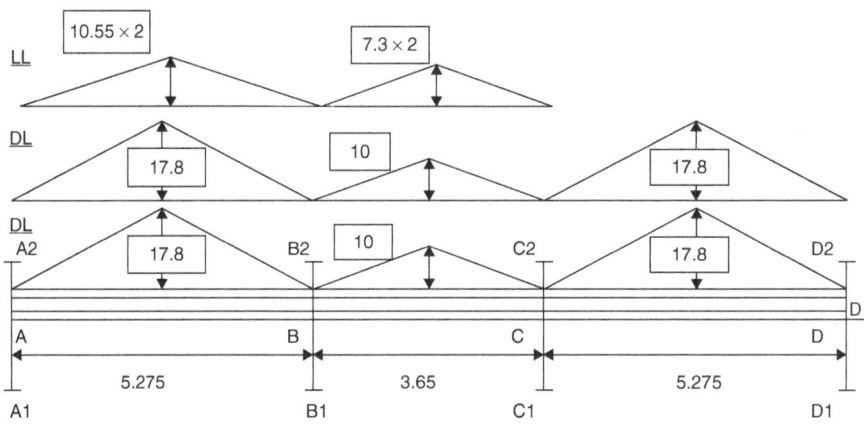

FIGURE 4.17 Load transferred from floor to substitute frame in the shorter direction (Figure 4.14) for getting maximum support moment at B

The live load/imposed load for all the slab panels = 4 kN/m²

Therefore, the maximum intensity of the triangular dead load on AB and CD = 2.63 x 6.8 = **17.8 kN/m**

(the loads from both sides of the beam from the slab panel are taken into account).

And the maximum intensity of the triangular dead load on BC = 1.82 x 5.55 = **10 kN/m** (again, the loads from both sides of the beam from the slab panel are taken into account). Similarly, the maximum intensity of the triangular live load on AB and CD = 2.63 x 4 = **10.55 kN/m²**

(with the loads from both sides of the beam from the slab panel considered). And the maximum intensity of the triangular live load on BC = 1.82 x 4 = **7.3 kN/m**

(with the loads from both sides of the beam from the slab panel considered).

The above-mentioned loadings are shown in Figure 4.17 to get the maximum bending moment at support B. The moment distribution method has been utilized, and the calculations are furnished in Table 4.26.

The distribution factors and fixed end moments are furnished in Tables 4.22 and 4.23. Expressions for calculating the fixed end moments are shown in Table 4.24; the fixed end moments are provided in Table 4.25.

TABLE 4.24 Fixed end moment for different type of load

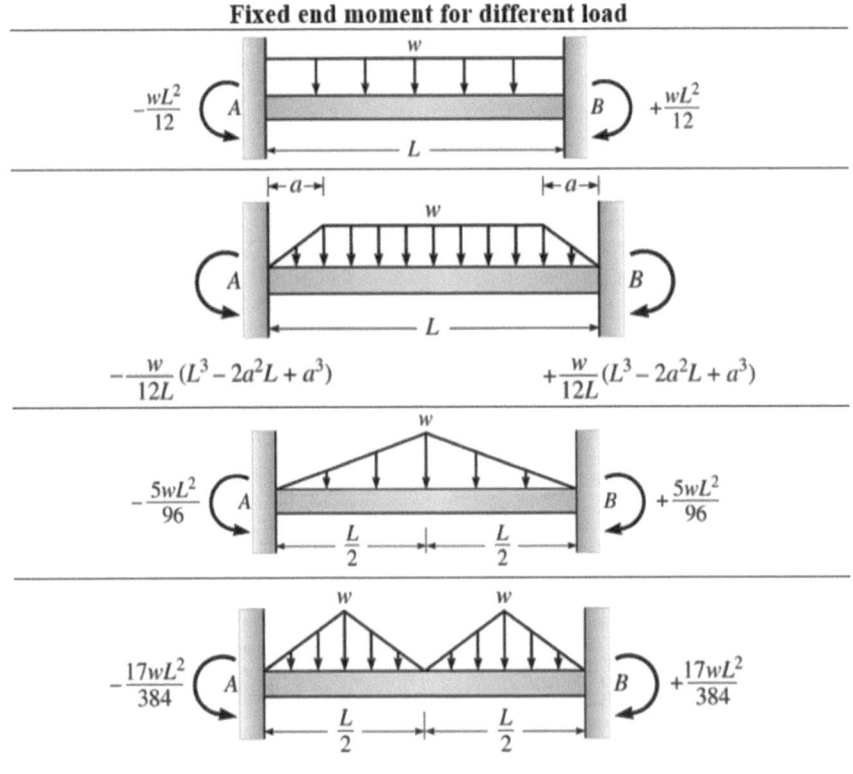

Fixed end moment for different load

$-\dfrac{wL^2}{12}$ A   B $+\dfrac{wL^2}{12}$

$-\dfrac{w}{12L}(L^3 - 2a^2L + a^3)$ $+\dfrac{w}{12L}(L^3 - 2a^2L + a^3)$

$-\dfrac{5wL^2}{96}$ A   B $+\dfrac{5wL^2}{96}$

$-\dfrac{17wL^2}{384}$ A   B $+\dfrac{17wL^2}{384}$

TABLE 4.25
Fixed end moments (see Figure 4.19a and Table 4.24)

Fixed end moment AB	(see Figure 4.19a)				
Uniform distributed load	$(6.71 + 3.43) \times (5.275)^2/12$		–/+	23.513	kNm
Triangular load	$(5 \times 17.8 \times (5.275)^2/96) \times 2$		–/+	51.593	kNm
Triangular load	$(5 \times 10.55 \times (5.275)^2/96) \times 2$		–/+	30.579	kNm
		Total	–/+	105.685	kNm
Fixed end moment BC					
Uniform distributed load	$(6.71 + 3.43) \times (3.65)^2/12$		–/+	11.258	kNm
Triangular load	$(5 \times 10 \times (3.65)^2/96) \times 2$		–/+	13.878	kNm
Triangular load	$(5 \times 7.3 \times (3.65)^2/96) \times 2$		–/+	10.13	kNm
		Total	–/+	35.266	kNm
Fixed end moment CD					
Uniform distributed load	$(6.71 + 3.43) \times (5.275)^2/12$		–/+	23.513	kNm
Triangular load	$(5 \times 17.8 \times (5.275)^2/96) \times 2$		–/+	51.593	kNm
		Total	–/+	75.106	kNm

TABLE 4.26
Moment distribution method applied to short frame 4–4 (see Figure 4.4)

Joint	A			B					C				D	
	AB	AA1	AA2	BA	BC	BB1	BB2	CB	CD	CC1	CC2	DC	DD1	DD2
Distribution factor	0.414	0.276	0.310	0.190	0.374	0.172	0.194	0.374	0.190	0.172	0.194	0.414	0.276	0.31
Fixed end moment	−105.685			105.685	−35.266			35.266	−75.106			75.106		
Balance	43.754	29.169	32.762	−13.380	−26.337	−12.112	−13.661	14.900	7.570	6.853	7.729	−31.094	−20.729	−23.283
Carried over	−6.690			21.877	7.450			−13.168	−15.547			3.785		
Balance	2.770	1.846	2.074	−5.572	−10.968	−5.044	−5.689	10.740	5.456	4.939	5.571	−1.567	−1.045	−1.173
Carried over	−2.786			1.385	5.370			−5.484	−0.783			2.728		
Balance	1.153	0.769	0.864	−1.283	−2.526	−1.162	−1.310	2.344	1.191	1.078	1.216	−1.129	−0.753	−0.846
Final bending moment (kNm)	−67.484	31.784	35.700	108.712	−62.277	−18.318	−20.661	44.597	−77.220	12.870	14.516	47.829	−22.527	−25.302

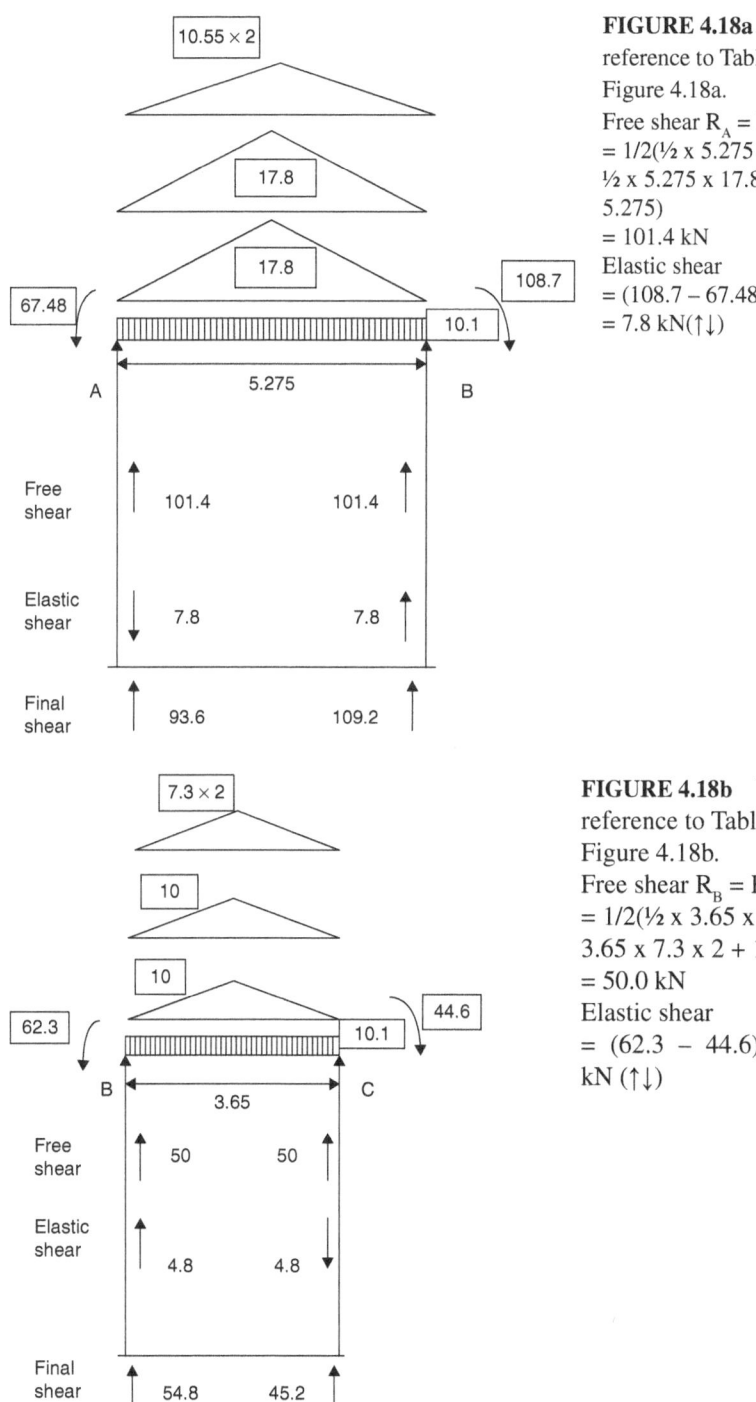

FIGURE 4.18a With reference to Table 4.7 and Figure 4.18a.

Free shear $R_A = R_B$
= 1/2(½ x 5.275 x 10.55 x 2+ ½ x 5.275 x 17.8 x 2+ 10.1 x 5.275)
= 101.4 kN

Elastic shear
= (108.7 – 67.48)/5.275
= 7.8 kN(↑↓)

FIGURE 4.18b With reference to Table 4.7 and Figure 4.18b.

Free shear $R_B = Rc$
= 1/2(½ x 3.65 x 10 x 2+ ½ x 3.65 x 7.3 x 2 + 10.1 x 3.65)
= 50.0 kN

Elastic shear
= (62.3 – 44.6)/3.65 = 4.8 kN (↑↓)

FIGURE 4.18 Free and elastic shear: (a) span AB and (b) span BC

Loading Case I

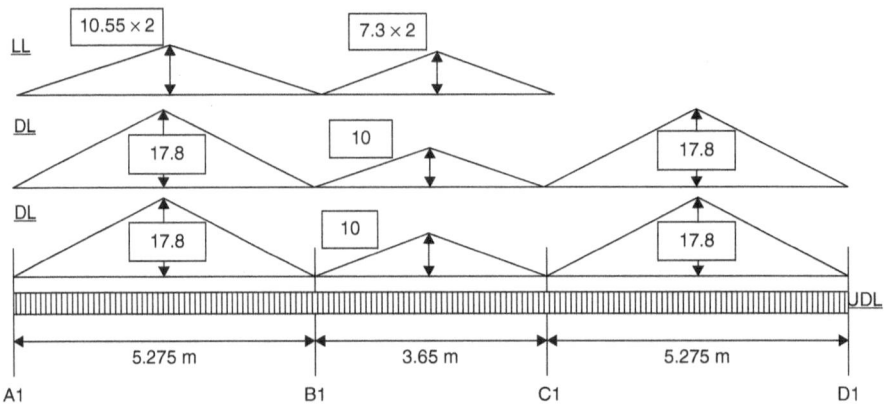

A1   B1   C1   D1

UDL = 6.71 kN/m (125 mm- thick brick wall) + 3.43 kN/m (self-weight of beam rib)

FIGURE 4.19a   Loading diagram for getting maximum support at B1

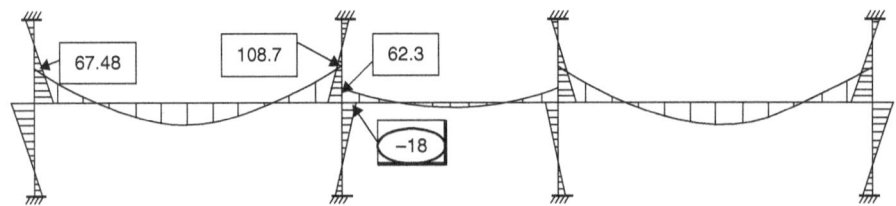

*Follow encircled value in later stages of calculations
*Moment of beam support B in kNm: M_{BA} = 108.7 kNm; M_{BC} = –62.3   kNm

FIGURE 4.19b   Bending moment diagram

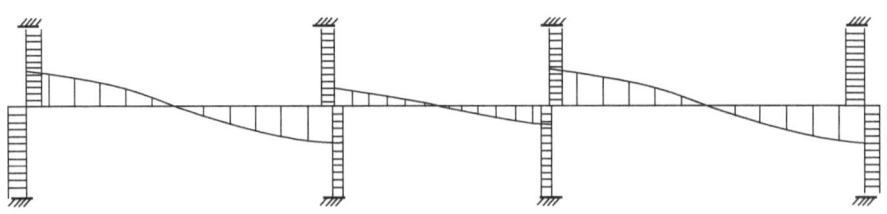

VA = + 93.6 kN; V_{BL} =–109.2; V_{BR} = + 54.8

FIGURE 4.19c   Shear force diagram

Loading Case II

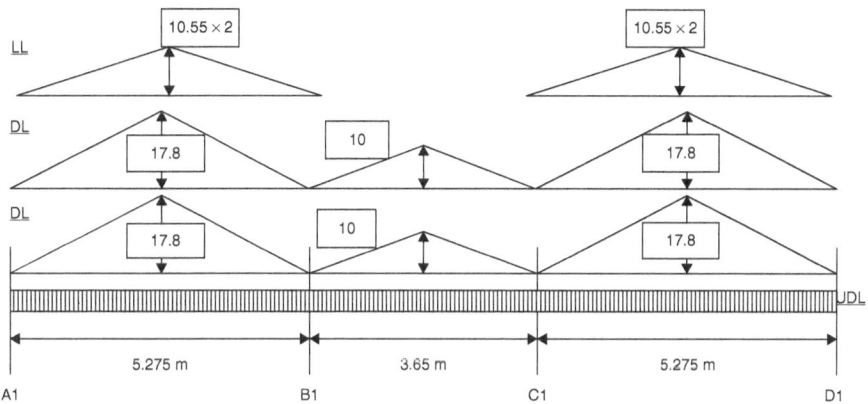

UDL = 6.71 kN/m (125 mm-thick brick wall) + 3.43 kN/m (self-weight of beam rib)

FIGURE 4.20a   Loading diagram for getting maximum span moment (AB)

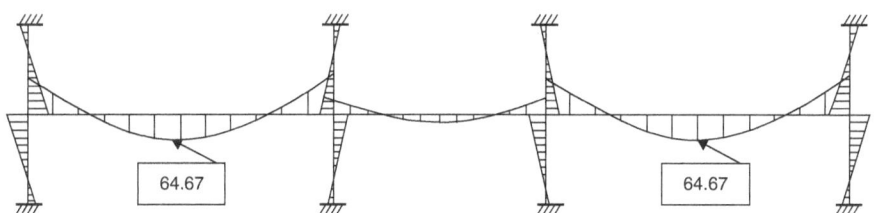

Span moment of beam AB and CD in kNm: M_{AB} = 64.67 kNm, M_{CD} = 64.67 kNm

FIGURE 4.20b   Bending moment diagram

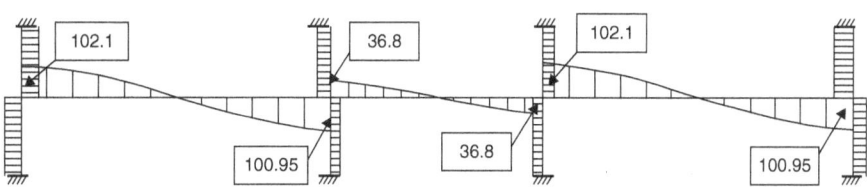

Beam shear at support in kN:
V_{AR} = +102.1 kN; V_{BL} = −100.95 kN; V_{BR} = +36.8 kN; V_{CL} = −36.8 kN;
V_{CR} = +102.1 kN; V_{DL} = −100.95 kN

FIGURE 4.20c   Shear force diagram

Loading Case III

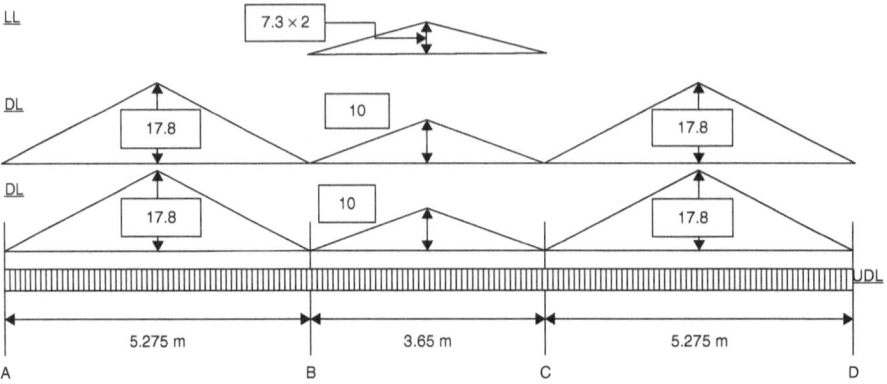

FIGURE 4.21a   Loading diagram for getting maximum span moment (BC)

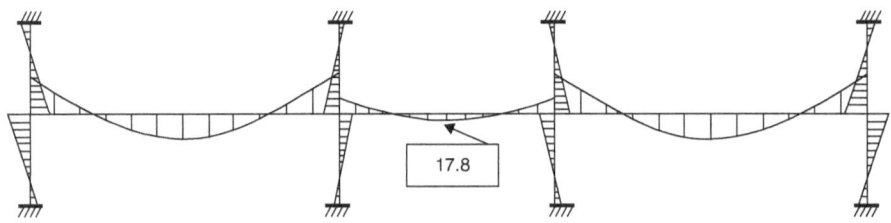

$M_{BC} = 17.8$ kNm

FIGURE 4.21b   Bending moment diagram

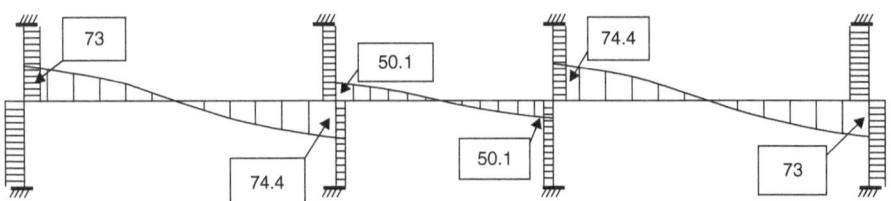

$V_{AR} = +73$ kN; $V_{BL} = -74.4$ kN; $V_{BR} = +50.1$ kN; $V_{CL} = -50.1$ kN, $V_{CR} = +74.4$ kN; $V_{DL} = -73$ kN

FIGURE 4.21c   Shear force diagram

Loading Case IV

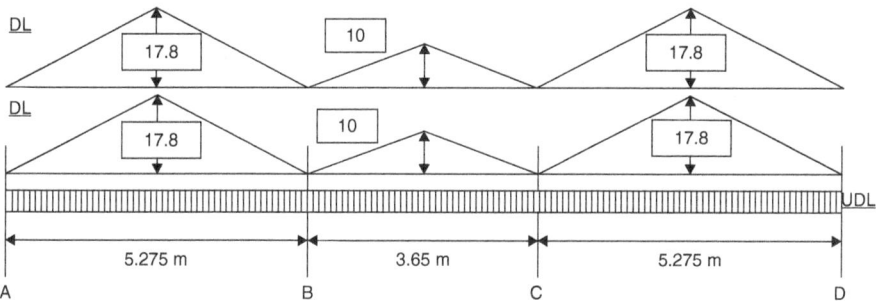

UDL = 6.71 kN/m (125 mm wall) + 3.43 kN/m (self-weight of beam rib)

FIGURE 4.22a   Loading diagram (Dead load only)

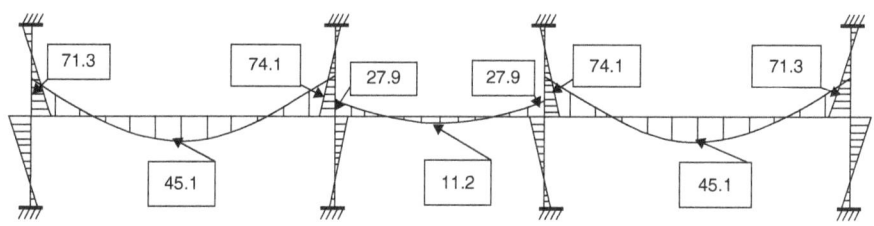

Moment of beam in kNm:

M_{AB} =–71.3 kNm; M_{BA} =–74.1; M_{BC} =–27.9; M_{CB} =–27.9; M_{CD} =–74.1; M_{DC} = –71.3
Span moment of beams AB, BC and CD in kNm: M_{AB} = 45. 1; M_{BC} = 11.2; M_{CD} = 45.1
*Follow encircled value in later stages of calculations

FIGURE 4.22b   Bending moment diagram

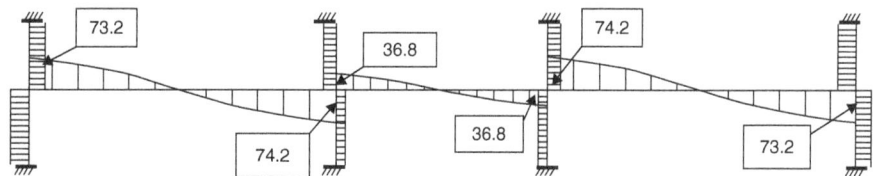

Beam shear at support in kN:

V_{AR} = +73.2 kN; V_{BL} = –74.2 kN; V_{BR} = +36.8 kN; V_{CL} = –36.8 kN; V_{CR} = +74.2 kN;
V_{DL} =–73.2 kN

FIGURE 4.22c   Shear force diagram

Examining the bending moment and shear force values, the design values due to the dead load and the live load are furnished in Tables 4.27, 4.28, 4.29 and 4.30 (referring to Figures 4.18,4.19, 4.20, 4.21 and 4.22).

TABLE 4.27
Design bending moment (kNm) due to dead load only

Support	A		B_{left}	B_{right}		C_{left}	C_{right}		D
Span		AB			BC			CD	
Support moment	−71.3		−74.1	−27.9		−27.9	−74.1		−71.3
Span moment		45.1			11.2			45.1	

TABLE 4.28
Design shear force (kN) due to dead load only

	A		B		C		D	
Support	Left	Right	Left	Right	Left	Right	Left	Right
Shear force		73.2	−74.2	36.8	−36.8	74.2	−73.2	

Note: Follow encircled values in later stages of calculations.

TABLE 4.29
Design bending moment (kNm) due to dead and live loads

Support	A		B_{left}	B_{right}		C_{left}	C_{right}		D
Span		AB			BC			CD	
Support moment	−67.5		−108.7	−62.3		−62.3	−108.7		−67.5
Span moment		64.67			17.8			64.67	

TABLE 4.30
Design shear force (kN) due to dead and live loads

	A		B		C		D	
Support	Left	Right	Left	Right	Left	Right	Left	Right
Shear force		102.1	−109.2	54.8	−54.8	109.2	−102.1	

Note: Follow encircled values in later stages of calculations.

The longer frame (1–2–3–4–5–6–7) B–B grid is as per Figure 4.15 and the dispersion of the load is as shown in Figure 4.14.

Column size = 700 mm x 300 mm and beam size = 650 mm x 250 mm as obtained tentatively beforehand.

The load dispersion as per the rule of IS 456 is followed, and it is shown in Figure 4.14.

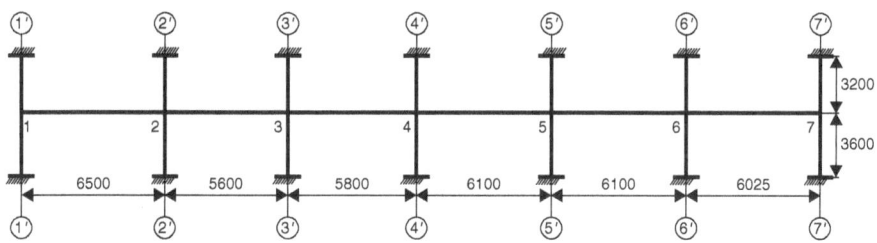

FIGURE 4.23 Substitute frame at first floor level in the longer direction (B–B)

TABLE 4.31
Moment of inertia of beams and columns of the longer frame (B–B) (see Figure 4.23)

	Width (b)	Depth (d)	Moment of inertia = $(1/12)\ bd^3$ (mm⁴)	Remarks
Beam	250	650	5,721,354,167	I_{B1}
Column	300	700	8,575,000,000	I_{C1}

TABLE 4.32
Stiffness of the beams and columns of the longer frame (see Figure 4.23)

Member	Length (mm)	Moment of inertia of beam (I_{B1}) (mm⁴)	Moment of inertia of column (I_{C1}) (mm⁴)	Stiffness (K)
1–2	6,500			$4EI_{B1}/6,500$
2–3	5,600			$4EI_{B1}/5,600$
3–4	5,800			$4EI_{B1}/5,800$
4–5	6,100			$4EI_{B1}/6,100$
5–6	6,100			$4EI_{B1}/6,100$
6–7	6,025	5,721,354,167	8,575,000,000	$4EI_{B1}/6,025$
1–1', 2–2', 3–3', 4–4', 5–5', 6–6', 7–7'	1,800			$4EI_{C1}/3,600$
1–1'', 2–2'', 3–3'', 4–4'', 5–5'', 6–6'', 7–7''	1,600			$4EI_{C1}/3,200$

TABLE 4.33
Distribution factor at the joints of the longer frame (see Figure 4.23)

Joint	Member	K	∑K	DF=$K_{ij}/\sum K$
1	1–2	3,520,833.333E		0.080
	1–1'	9,527,777.778E	23,767,361.11E	0.010
	1–1"	10,718,750E		0.451
2	2–1	3,520,833.333E		0.126
	2–3	4,086,681.548E	27,854,042.66E	0.147
	2–2'	9,527,777.778E		0.342
	2–2"	10,718,750E		0.385
3	3–2	4,086,681.548E		0.145
	3–4	3,945,761.494E	28,278,970.82E	0.140
	3–3'	9,527,777.778E		0.337
	3–3"	10,718,750E		0.379
4	4–3	3,945,761.494E		0.141
	4–5	3,751,707.650E	27,943,996.92E	0.134
	4–4'	9,527,777.778E		0.341
	4–4"	10,718,750E		0.384
5	5–4	3,751,707.650E		0.135
	5–6	3,751,707.650E	27,749,943.08E	0.135
	5–5'	9,527,777.778E		0.343
	5–5"	10,718,750E		0.386
6	6–5	3,751,707.650E		0.135
	6–7	3,798,409.405E	27,796,644.83E	0.137
	6–6'	9,527,777.778E		0.343
	6–6"	10,718,750E		0.386
7	7–6	3,798,409.405E		0.158
	7–7'	9,527,777.778E	24,044,937.18E	0.396
	7–7"	10,718,750E		0.446

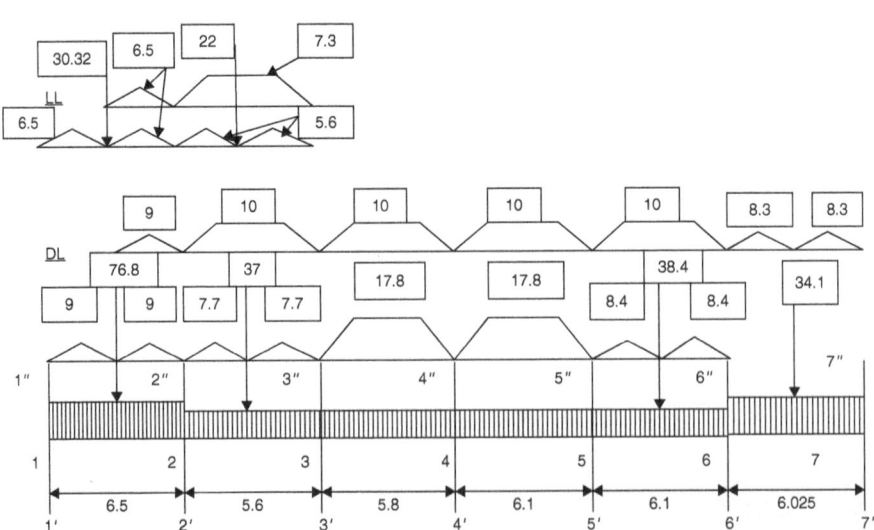

FIGURE 4.24   Loading diagram of the substitute frame in the longer direction (Figure 4.14) to get maximum support moment at 2

UDL = 6.71 kN/m (125th Wall)/ 13.43 kN/m (250th Wall) + 3.43 kN/m (Self weight of beam rib)

For 1–2, 2–3, 5–6 and 6–7 the load intensity (w) = 5.5 kN/m², and for 3–4 and 4–5 the load intensity (w) = 6.8 kN/m² (refer to the load calculation).
The live load/imposed load for all the slab panels = 4 kN/m²
Therefore, the maximum intensity of the triangular dead load on 1–2

$$= 3.25/2 \text{ x } 5.5 = \textbf{8.93} ≈\textbf{9 kN/m}$$

(the loads from both sides of the beam from the slab panel are taken into account).
The point load from the cross secondary beam at the mid-point of 1–2 = 76.8 kN
The maximum intensity of the triangular dead load on 2–3

$$= 2.8/2 \text{ x } 5.5 = \textbf{7.7 kN/m}$$

(again, the loads from both sides of the beam from the slab panel are taken into account).
The maximum intensity of the trapezoidal dead load on 2–3

$$= 3.65/2 \text{ x } 5.5 = \textbf{10 kN/m}$$

The point load from the cross secondary beam at the mid-point of 2–3 = 37 kN
The maximum intensity of the trapezoidal dead load on 3–4 and 4–5

$$= 3.65/2 \text{ x } 5.5 = 10 \text{ kN/m}$$

The maximum intensity of the trapezoidal dead load on 3–4 and 4–5 = 5.275/2 x 6.8 = **17.8** kN/m

The maximum intensity of the triangular dead load on 5–6

$$= 3.05/2 \text{ x } 5.5 = \textbf{8.4 kN/m}$$

(with the loads from both sides of the beam from the slab panel taken into account).
The maximum intensity of the trapezoidal dead load on 5–6

$$= 3.65/2 \text{ x } 5.5 = \textbf{10 kN/m}$$

The point load from the cross secondary beam at the mid-point of 5–6 = 38.4 kN
The maximum intensity of the triangular dead load on 6–7

$$= 3.013/2 \text{ x } 5.5 = \textbf{8.3 kN/m}$$

(with the loads from both sides of the beam from the slab panel taken into account). The point load from the cross secondary beam at the midpoint of 6–7 = 34.1 kN Similarly, the maximum intensity of the triangular live load on 1–2

$$= 3.25/2 \times 4 = 6.5 \text{ kN/m}$$

(with the loads from both sides of the beam from the slab panel taken into account). The point load from the cross secondary beam at the midpoint of 1–2 = 30.32 kN The maximum intensity of the triangular dead load on 2–3

$$= 2.8/2 \times 4 = 5.6 \text{ kN/m}$$

(with the loads from both sides of the beam from the slab panel taken into account). The maximum intensity of the trapezoidal dead load on 2–3

$$= 3.65/2 \times 4 = 7.3 \text{ kN/m}$$

The point load from the cross secondary beam at the midpoint of 2–3 = 22 kN

The above-mentioned loadings are shown in Figure 4.24 to get maximum bending moment at support 2. The fixed end moments and moment distribution of the longer frame are furnished in Tables 4.34 and 4.35.

TABLE 4.34
Fixed end moments (see Figure 4.24)

Fixed end moment for span 1–2			**Refer to Figure 4.25a**			
UDL		–/+	59.36	kNm		
Triangular load (right to point load)		–/+	16.83	kNm		
Triangular load (right to point load)		–/+	12.16	kNm		
Triangular load (left to point load)		–/+	4.97	kNm		
		–/+			2.535	kNm
Triangular load (left to point load)		–/+	6.89			
		–/+			3.510	kNm
Point load		–/+	87.04	kNm		
	Total AT2	(+)	187.25	AT 1 (–)	181.435	kNm
Fixed end momentfor span2–3						
UDL		–/+	26.50	kNm		
Triangular load (right to point load)		–/+	10.69	kNm		
Triangular load (right to point load)		–/+	7.77	kNm		
Trapezoidal		(–)	9.72			
					9.720	
Trapezoidal		(–)	7.10			
					7.100	
Point load		–/+	41.30			
	Total	(–)	103.08		103.080	kNm

(continued)

TABLE 4.34 (Continued)
Fixed end moments (see Figure 4.24)

Fixed end moment for span 1–2			Refer to Figure 4.25a			
Fixed end moment for span 3–4						
UDL			–/+	28.43	kNm	
Trapezoidal			(–)	19.69		
						19.690
Trapezoidal			(–)	11.06		
						11.060
	Total		(–)	59.18		59.180 kNm
Fixed end moment for span 4–5						
UDL			–/+	31.44	kNm	
Trapezoidal			(–)	23.21		
						23.210
Trapezoidal			(–)	13.04		
						13.040
	Total		(–)	67.69		67.690
Fixed end moment for span 5–6						
UDL			–/+	31.44	kNm	
Triangular load (right to point load)			–/+	13.84	kNm	
Trapezoidal			(–)	12.40		
						12.400
Point load			–/+	44.53	kNm	
	Total			102.21		102.210 kNm
Fixed end moment for span 6–7						
UDL			–/+	52.28	kNm	
Triangular load (right to point load)			–/+	13.34	kNm	
Point load			–/+	25.68	kNm	
	Total			91.30	kNm	

Moment distribution method is applied and calculations are furnished in Tables 4.35, considering distribution factor and fixed end moment furnished in Tables 4.33 and 4.34

These data are applied and the calculations are furnished in Table 4.35, taking into account the distribution factors and fixed end moments as furnished in Tables 4.33 and 4.34.

TABLE 4.35
Moment distribution method applied for analysis of the long frame (B–B)

Joint	1-2	1-1'	1-1"	2-1	2-3	2-2'	2-2"	3-2	3-4	3-3'	3-3"	4-3	4-5	4-4'	4-4"	5-4	5-6	5-5'	5-5"	6-5	6-7	6-6'	6-6"	7-6	7-7'	7-7"
	1			2				3				4				5				6				7		
Distribution factor	0.148	0.401	0.451	0.126	0.147	0.342	0.385	0.145	0.14	0.337	0.379	0.141	0.134	0.341	0.384	0.135	0.135	0.343	0.386	0.135	0.137	0.343	0.386	0.158	0.396	0.446
Fixed end moment (kNm)	-181.4			187.3	-103.1			103.1	-59.2			59.2	-67.7			67.7	-102.2			102.2	-91.3			91.3		
Balance	26.9	72.8	81.8	-10.6	-12.4	-28.8	-32.4	-6.4	-6.1	-14.8	-16.6	1.2	1.1	2.9	3.3	4.7	4.7	11.8	13.3	-1.5	-1.5	-3.7	-4.2	-14.4	-36.2	-40.7
Carried over	-5.3			13.4	-3.2			-6.2	-7.2			-3.1	0.0			0.6	0.0			2.3	0.0			-3.1		
Balance	0.8	2.1	2.4	-1.3	-1.5	-3.5	-3.9	1.9	1.9	4.5	5.1	0.4	0.4	1.0	1.2	-0.1	-0.1	-0.2	-0.2	-0.3	-0.3	-0.8	-0.9	0.5	1.2	1.4
Carried over	-0.6			0.4	1.0			-0.8	0.2			0.9	0.0			0.2	0.0			0.0	0.0			0.9		
Balance	0.1	0.3	0.3	-0.2	-0.2	-0.5	-0.5	0.1	0.1	0.2	0.2	-0.1	-0.1	-0.3	-0.4	0.0	0.0	-0.1	-0.1	0.0	0.0	0.0	0.0	-0.1	-0.4	-0.4
Final moment (kNm)	-159.7	75.1	84.5	189.0	-119.4	-32.8	36.9	91.8	-70.3	-10.1	-11.4	58.5	-66.3	3.6	4.1	73.0	-97.7	11.6	13.0	102.7	-93.1	-4.5	-5.1	75.1	-35.3	-39.8

Loading Case I

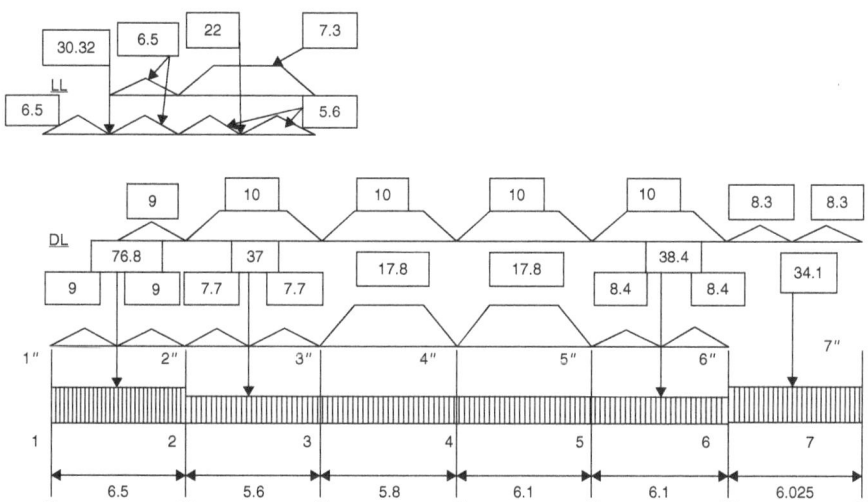

UDL = 6.71 kN/m (125 mm brick wall)/13.43 kN/m (250 mm brick wall) + 3.43 kN/m
(self-weight of beam rib)

FIGURE 4.25a   Loading diagram for getting maximum support moment at 2

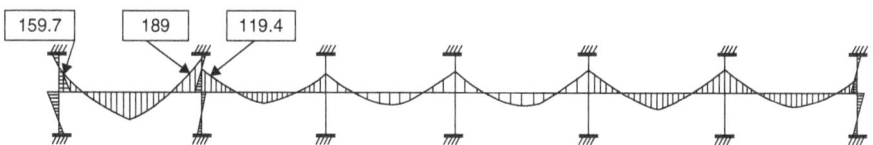

M_{12} = 159.7 kNm; M_{21} = 189 kNm; M_{23} = 119.4 kNm

FIGURE 4.25b   Bending moment diagram

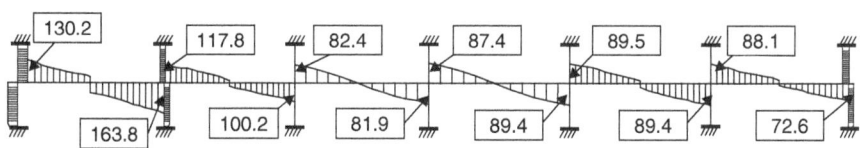

Beam shear at supports in kN

FIGURE 4.25c   Shear force diagram

Loading Case II

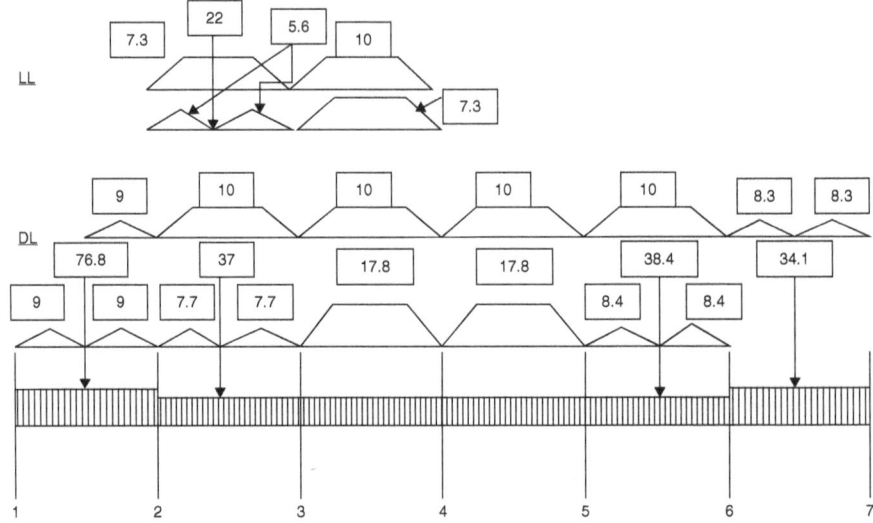

UDL = 6.71 kN/m (125 mm brick wall)/13.43 kN/m (250 mm brick wall) + 3.43 kN/m (self-weight of beam rib)

Note: Point load in kN, distributed load in kN/m, length in m.

FIGURE 4.26a   Loading diagram for getting maximum support moment at 3

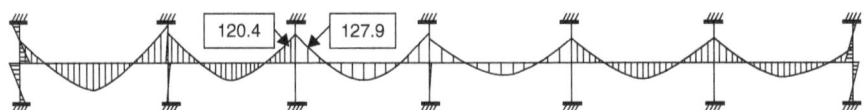

Moment of beam support 3 in kNm: M_{32} = 120.42 kNm/M_{34} = 127.93 kNm

FIGURE 4.26b   Bending moment diagram

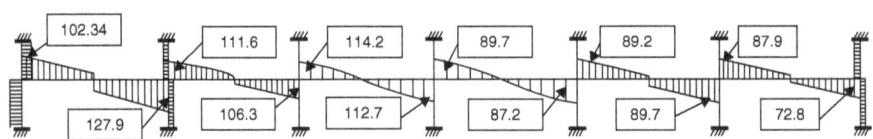

Beam shear at supports in kN.

FIGURE 4.26c   Shear force diagram

Loading Case III

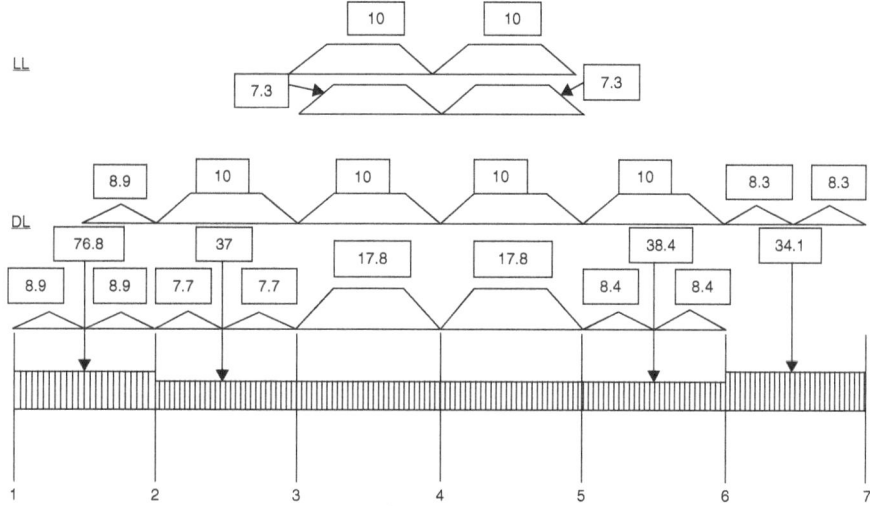

UDL = 6.71 kN/m (125 mm-thick brick wall)/13.43 kN/m (250 mm-thick wall) + 3.43 kN/m (self-weight of beam rib)

Note: Point load in kN, distributed load in kN/m, length in m.

FIGURE 4.27a Loading diagram to get maximum support moment at 4

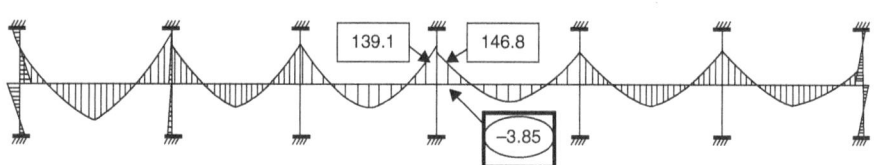

*Follow encircled value in later stages of calculations

Moment of beam support 4 in kNm: M_{43} = 139.1 kNm/M_{45} = 146.8 kNm

FIGURE 4.27b Bending moment diagram

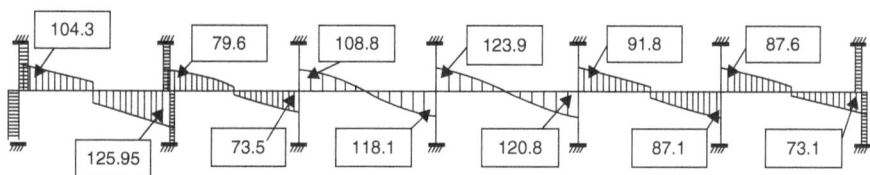

Beam shear at supports in kN.

FIGURE 4.27c Shear force diagram

Loading Case IV

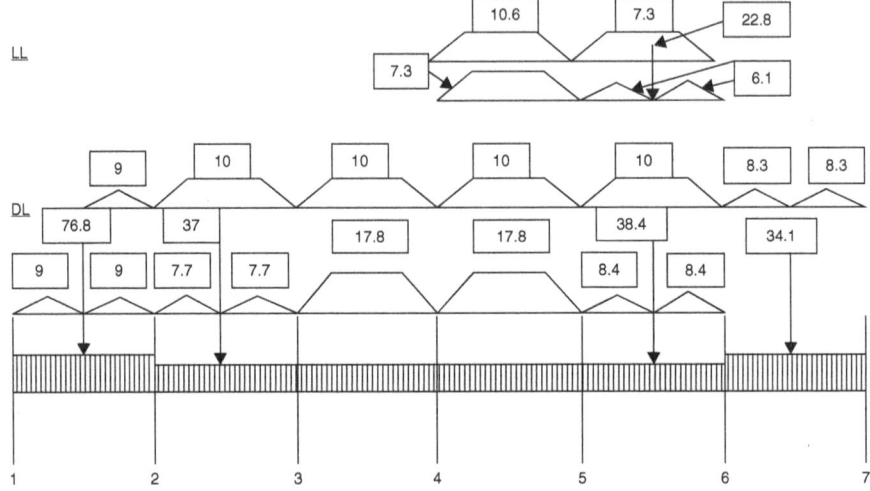

UDL = 6.71 kN/m (125 mm wall)/13.43 kN/m (250 mm wall) + 3.43 kN/m
(self-weight of beam rib)
Note: Point load in kN, distributed load in kN/m, length in m.

FIGURE 4.28a   Loading diagram to get maximum support at 5

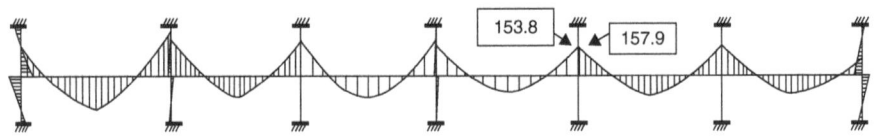

Moment of beam support 5 in kNm: M_{54} = 153.8 kNm/M_{56} = 157.9 kNm

FIGURE 4.28b   Bending moment diagram

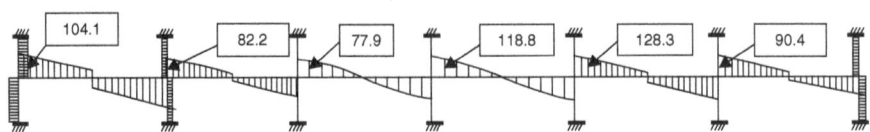

Beam shear at supports in kN.

FIGURE 4.28c   Shear force diagram

Loading Case V

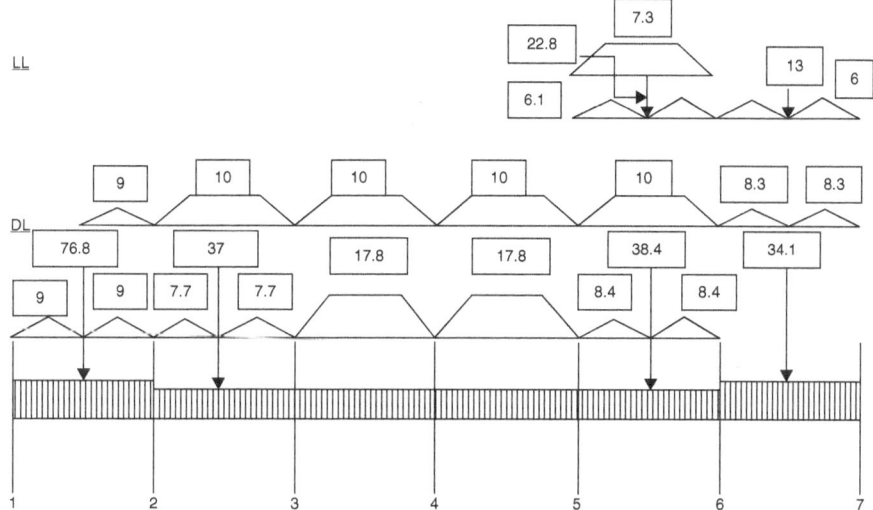

UDL = 6.71 kN/m (125 mm wall)/13.43 kN/m (250 mm wall) + 3.43 kN/m (self-weight of beam rib)

Note: Point load in kN, Distributed load in kN/m, Length in m

FIGURE 4.29a Loading diagram to get maximum support moment at 6

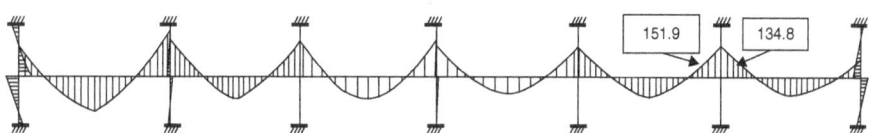

Moment of beam support 6 in kNm: M_{65} = 151.9 kNm/M_{67} = 134.8 kNm

FIGURE 4.29b Bending moment diagram

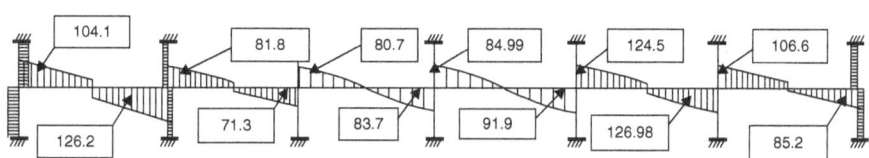

Beam shear at supports in kN.

FIGURE 4.29c Shear force diagram

Loading Case VI

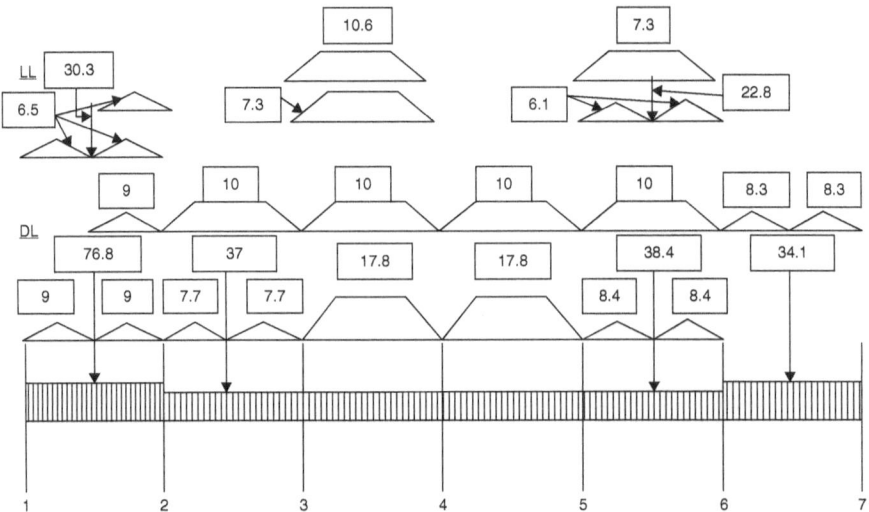

UDL = 6.71 kN/m (125 mm-thick wall)/13.43 kN/m (250 mm-thick wall) +
3.43 kN/m (self-weight of beam rib)
Note: Point load in kN, distributed load in kN/m, length in m.

FIGURE 4.30a   Loading diagram for getting maximum span moment (1-2, 3-4, 5-6)

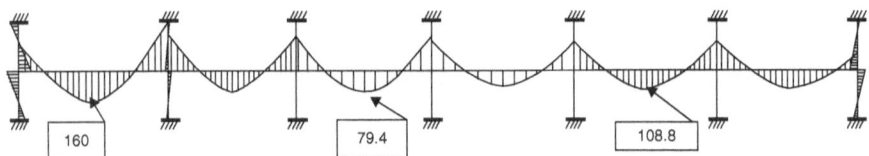

Span moments of beams 1–2, 3–4 and 5–6 in kNm:
M_{1-2} = 160 kNm; M_{3-4} = 79.4 kNm; M_{5-6} = 108.8 kNm

FIGURE 4.30b   Bending moment diagram

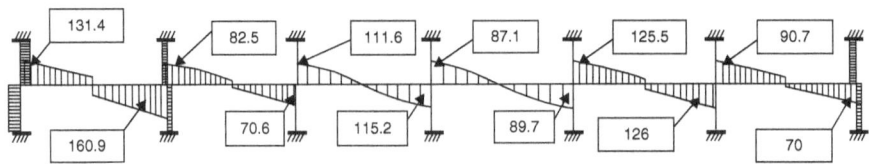

Beam shear at supports in kN.

FIGURE 4.30c   Shear force diagram

Loading Case VII

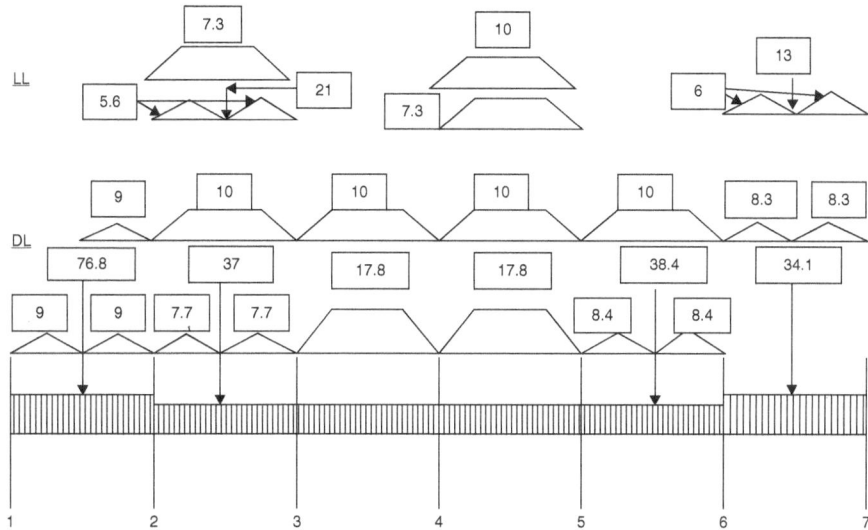

UDL = 6.71 kN/m (125 mm-thick wall)/13.43 kN/m (250 mm-thick wall) +
3.43 kN/m (self-weight of beam rib)
Note: Point load in kN, distributed load in kN/m, length in m.

FIGURE 4.31a   Loading diagram for maximum span moment (2-3, 4-5, 6-7)

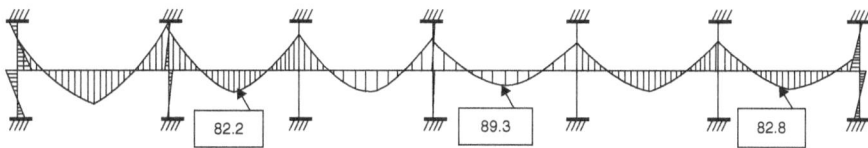

Span moments of beams 2–3, 4–5 and 6–7 in kNm:
$M_{2-3} = 82.2$ kNm; $M_{4-5} = 89.3$ kNm; $M_{6-7} = 82.8$ kNm

FIGURE 4.31b   Bending moment diagram

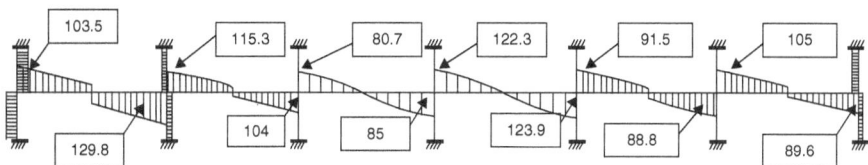

Beam shear at supports in kN.

FIGURE 4.31   Shear force diagram

Loading Case VIII

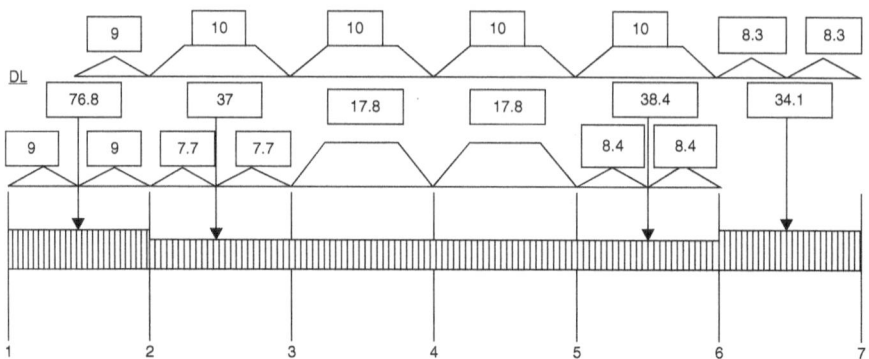

UDL = 6.71 kN/m (125 mm-thick wall)/13.43 kN/m (250 mm-thick wall) + 3.43 kN/m (self-weight of beam rib)
Note: Point load in kN, distributed load in kN/m, length in m.

FIGURE 4.32a Loading diagram for dead load only

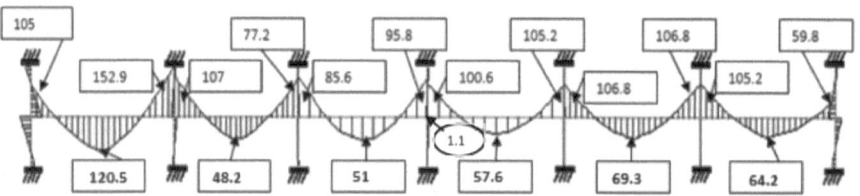

Moment of Beam supports in kNm: M_{12} = 105kNm M_{21} = 152.9 M_{23} = 107 M_{32} = 77.2
M_{34} = 85.6 M_{43} = 95.8 M_{45} = 100.6 M_{54} = 105.2 M_{56} = 106.8 M_{65} = 106.8 M_{67} = 105.2 and
M_{76} = 59.8kNm
Span moments of beam in kNm: M_{1-2} = 120.5kNm, M_{2-3} = 48.2kNm, M_{3-4} = 51kNm
M_{4-5} = 57.6kNm, M_{5-6} = 69.3 kNm, and M_{6-7} = 64.2kNm

FIGURE 4.32b Bending moment diagram

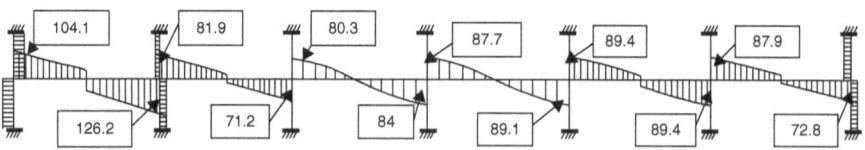

Beam shear at supports in kN

FIGURE 4.32c Shear force diagram

Examining the bending moments and shear force values, the design values due to the dead load and the live load are furnished in Tables 4.36, 4.37, 4.38 and 4.39 (with reference to Figures 4.25, 4.26, 4.27, 4.28, 4.29, 4.30, 4.31 and 4.32).

TABLE 4.36
Design bending moment (kNm) due to dead load only

Support	Right		Left	Right		Left	Right		Left	Right	
	1		2			3			4		
Span		1–2			2–3			3–4			4–5
Support moment	–105		–152.9	–107		–77.2	–85.6		–95.8	–100.6	
Span moment		120.5			48.2			51			57.6
Support		5			6			7				
	Left	Right		Left	Right		Left				
Span		5–6			6–7						
Support moment	–105.2	–106.8		–106.8	–105.2		–59.8				
Span moment		69.3			64.2						

Note: Follow encircled value in later stages of calculations.

TABLE 4.37
Design shear force (kN) due to dead load only

Support	Left	Right	Left	Right	Left	Right	Left	Right	Left	Right	Left	Right	Left
	1		2		3		4		5		6		7
Shear force		104.1	–126.2	81.9	–71.2	80.3	(–84)	(87.7)	–89.1	89.4	–89.4	87.9	–72.8

TABLE 4.38
Design bending moment (kNm) due to dead and live loads

Support	Right		Left	Right		Left	Right		Left	Right	
	1		2			3			4		
Span		1–2			2-3			3-4			4-5
Support moment	–159.7		–189	–119.4		–120.4	-127.9		–139.1	–146.8	
Span moment		160			82.2			79.4			89.3
Support	5			6			7				
	Left	Right		Left	Right		Left				
Span		5–6			6-7						
Support moment	–153.8	– 157.9		–151.9	-134.8		-70.3				
Span moment		108.8			82.8						

TABLE 4.39
Design shear force (kN) due to dead and live loads

	1		2		3		4		5		6		7
Support	**Left**	**Right**	**Left**	**Right**	**Left**	**Right**	**Left**	**Right**	**Left**	**Right**	**Left**	**Right**	**Left**
Shear force		130.2	–163.8	117.8	–106.3	114.2	⊖118.1	⊙124	–125.9	128.3	–127.0	106.6	–85.2

Note: Follow encircled value in later stages of calculations.

4.4.10 FRAME ANALYSIS UNDER WIND AND SEISMIC FORCES

For any building frame (considered as a two-dimensional frame) subjected to horizontal loads such as wind or seismic loads, there are in general two approximate methods based on assumptions that can be applied to solve these statically indeterminate frames: use of the portal method or the cantilever method. It has been observed that the cantilever method is more popular and gives results close to rigorous or computer-aided analysis for more or less regular frames. As per clause 22.4.3 of IS 456, 2000, simplified methods may be used to obtain the moments and shears against lateral loads for symmetrical structures. For unsymmetrical structures, more rigorous methods should be used. This building is more or less symmetrical. Therefore, the cantilever method can be applied without sacrificing accuracy to any great extent.

It is assumed that the whole frame will deflect laterally in the similar way to a vertical cantilever. A stress diagram of a cantilever frame under lateral force is shown in Figure 4.33.

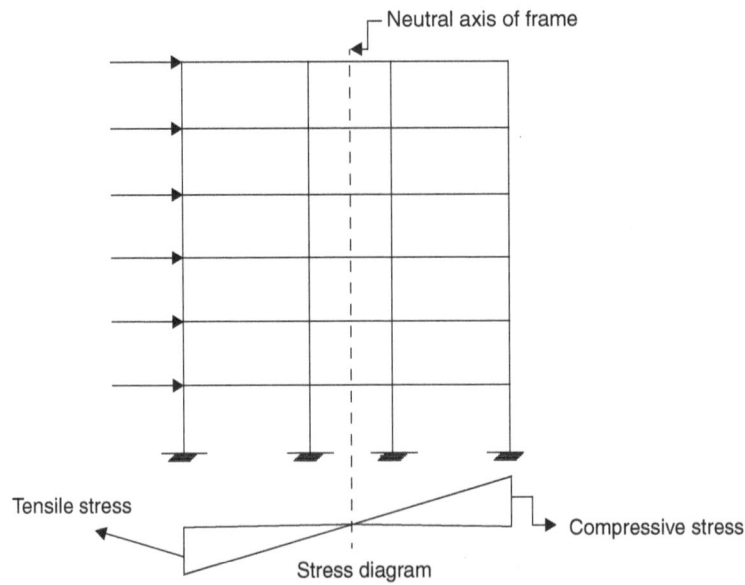

FIGURE 4.33 Stress diagram for cantilever under horizontal loading

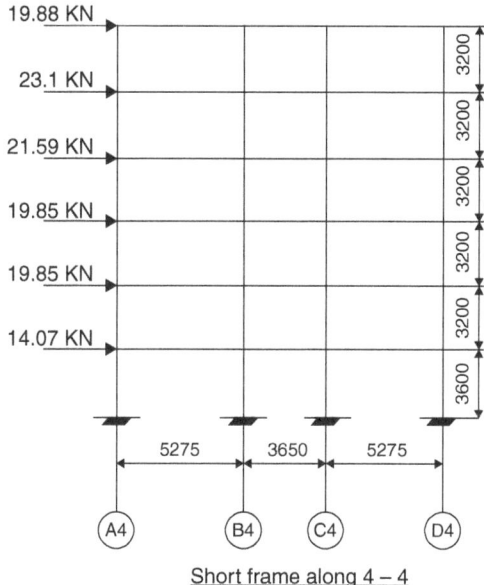

FIGURE 4.34 Frame in the shorter direction (4–4) showing wind loads

The location of the neutral axis of the whole frame needs to be calculated by considering the cross-sectional areas and locations of the columns.

The following assumptions are adopted in the cantilever method.

1. The point of contra flexure (the inflection point) will occur at the mid-point of all the horizontal and vertical members of the frame.
2. The direct stresses in each column are proportional to the distance of that column from the center of gravity of the frame.

The following formula has been used to locate neutral axis of the frame.

$$X' = \sum A_i x_i / \sum A$$

where

A = column area of cross-section = 0.7 m x 0.7 m = 0.21 m²
(assuming the column size is the same throughout the height)

$$X' \text{ from A} = \frac{A (0 + 5.275 + (5.275 + 3.65) + (5.275 + 3.65))}{4A}$$

Column load = Axial stress (Δ) x Area of column

Let us apply the cantilever method in connection with the frame shown in Figure 4.34; detailed calculations are shown in Table 4.40.

TABLE 4.40
Cantilever method of analysis

Free body diagram (FBD) of sixth floor
The wind force is acting on the longer face of the building

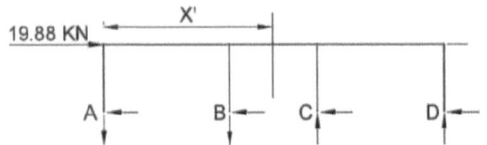

Let, $\Delta 1$ is stress in left column at A
So, Stress in 2nd column at B
$(7.1–5.275)/7.175$ x $\Delta 1 = 0.25436$
$\Delta 1$, direct stress of a column is
proportional to the distance of
that particular column from CG
of frame. Similarly, Stress in 3rd
column at C $(1.825/7.175)$ x $\Delta 1$
$= 0.25436 \Delta 1$ Stress in 4th column
at D $7.1/7.1$ x $\Delta 1 = 1 \Delta 1 \Sigma M = 0$
Taking moment about D 19.88 x
$1.6 – \Delta 1 x A x 14.2–0.25436 \Delta 1$ x
A x$(5.275+3.65)$ $+0.25436$ x $\Delta 1$x
$5.275 = 0$ Column load $\Delta 1$ x A=
2.09936 kN at A and D Column
load 0.53399 kN at B and C

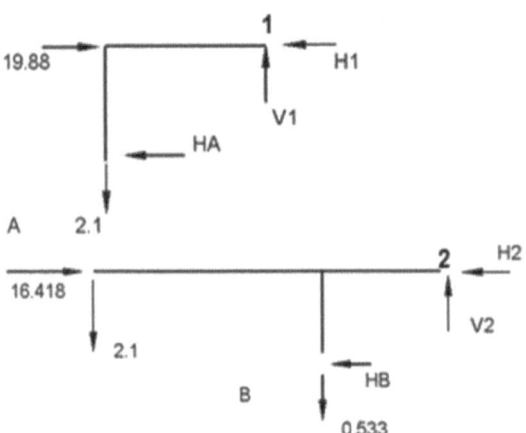

$\Sigma V = 0$
$V1 = 2.1$ kN
Taking the moment about 1:
$HA = 3.4617$ kN
$\Sigma H = 0$
$H1 = 16.418$ kN

$\Sigma V = 0$
$V2 = 2.633$ kN
Taking the moment about 2:
$HB = 6.465$ kN
$\Sigma H = 0$
$H2 = 9.953$ kN

(*continued*)

TABLE 4.40 (Continued)
Cantilever method of analysis

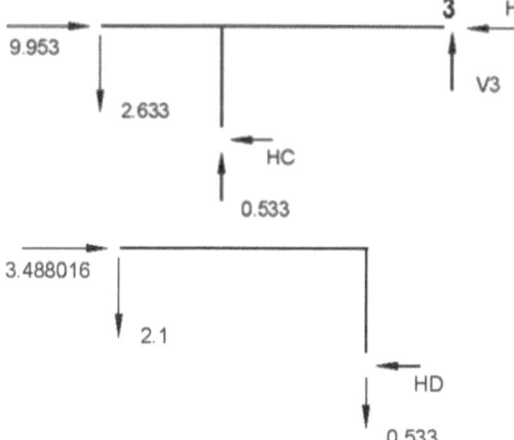

$\sum V = 0$
V3 = 2.1 kN
Taking the moment about 3:
HC = 6.465 kN
$\sum H = 0$
H3 = 3.488 kN

$\sum H = 0$
HD = 3.488 kN

FBD of fifth floor

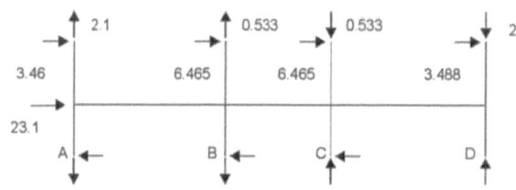

$\sum M = 0$
Taking moment about D
23.1 x 1.6 + (3.46 + 6.465 +
6.465 + 3.488)*3.2 + 2.1*14.2 +
0.533*8.925 – 0.533*5.275 –
Δ2 x A x 14.2 – 0.25436
Δ2 x A x (5.275 + 3.65)
+ 0.25436 x Δ2 x 5.275 = 0
Column load Δ2 x A = 8.737 kN
at A and D Column load = 2.222
kN at B and C

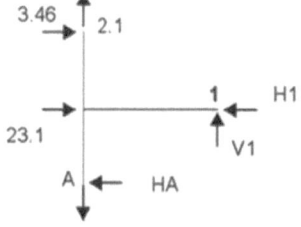

$\sum V = 0$
V1 = 6.637 kN
Taking the moment about 1:
H_A = 7.4797 kN
$\sum H = 0$
H1 = 19.082 kN.

TABLE 4.40 (Continued)
Cantilever method of analysis

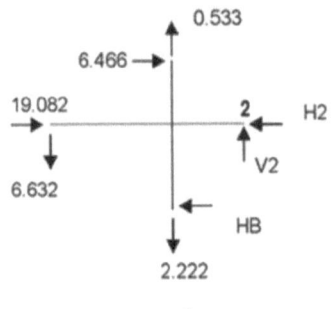

$\Sigma V = 0$
$V2 = 8.326$ kN
Taking the moment about 2:
$H_B = 13.973$ kN
$\Sigma H = 0$
$H2 = 11.574$ kN.

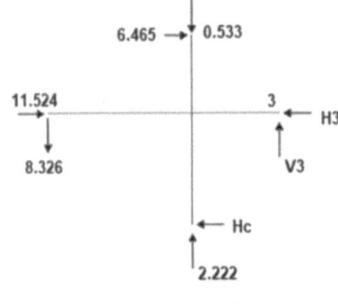

$\Sigma V = 0$
$V3 = 6.637$ kN
Taking the moment about 3:
$Hc = 13.973$ kN
$\Sigma H = 0$
$H3 = 4.067$ kN.

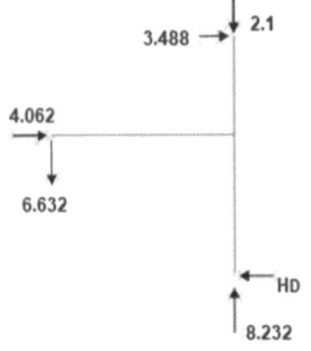

$\Sigma H = 0$
$HD = 7.555$ kN.

FBD of fourth floor

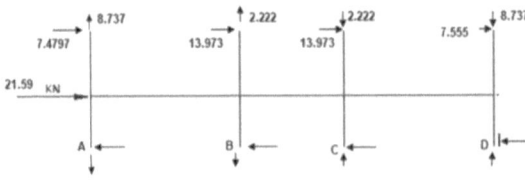

$\Sigma M = 0$
Taking moment about D
$21.59 \times 1.6 + (7.4797 + 13.973$
$+ 13.973 + 7.555) \times 3.2 +$
$8.737 \times 14.2 + 2.222 \times 8.925$
$- 2.222 \times 5.275 - \Delta3 \times A \times 14.2$
$- 0.25436\ \Delta3 \times A \times (5.275$
$+ 3.65) + 0.25436 \times \Delta3 \times 5.275$
$= 0$ Column load $\Delta3 = 20.112$ kN

(continued)

TABLE 4.40 (Continued)
Cantilever method of analysis

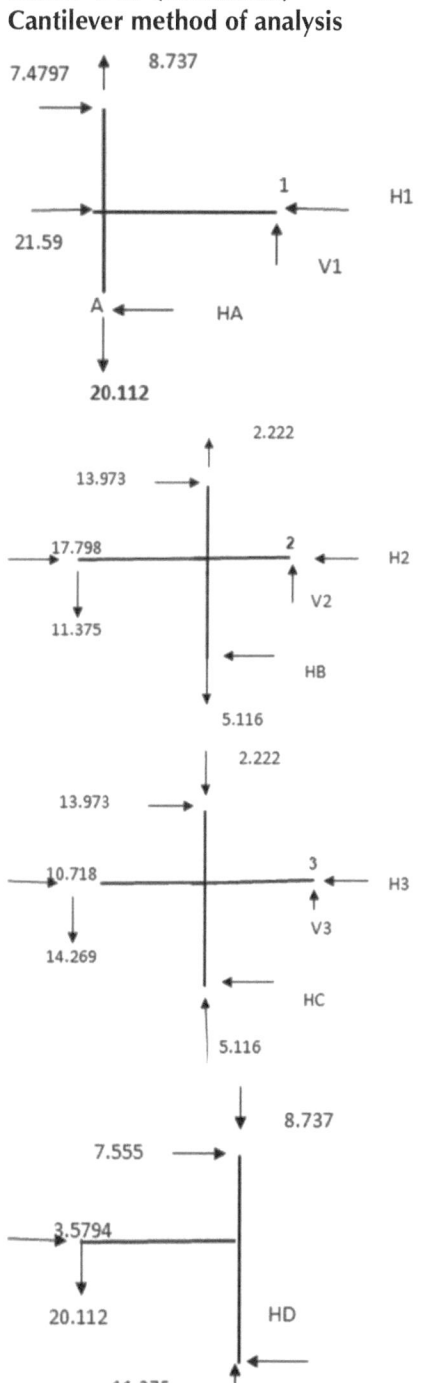

$\Sigma V = 0$
V1 = 1.375   kN
Taking the moment about 1:
HA = 11.271 kN
$\Sigma H = 0$
H1 = 17.798 kN

$\Sigma V = 0$
V2 = 14.269 kN
Taking the moment about 2:
HB = 21.054 kN
$\Sigma H = 0$
H2 = 10.718 kN

$\Sigma V = 0$
V3 = 11.375 kN
Taking the moment about 3:
Hc = 21.052 kN
$\Sigma H = 0$
H3 = 3.5794 kN

$\Sigma H = 0$
HD = 11.1344 kN

TABLE 4.40 (Continued)
Cantilever method of analysis

FBD of third floor

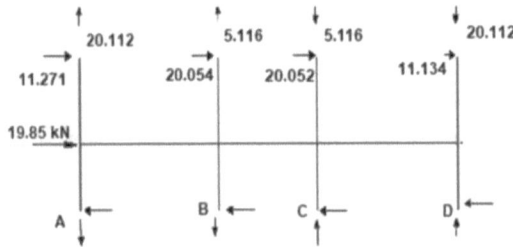

$\Sigma M = 0$
Taking the moment about D
19.85 x 1.6 + (11.271 + 21.054
+ 21.052 + 11.134)*3.2
+ 20.112*14.2 + 5.116*8.925 −
5.116*5.275 − Δ2 x A x 14.2 −
0.25436 Δ2 x A x (5.275 + 3.65)
+ 0.25436 x Δ2 x 5.275 = 0.
Column load Δ2 x A = 35.857 kN
at A & D = 9.121kN at B and C.

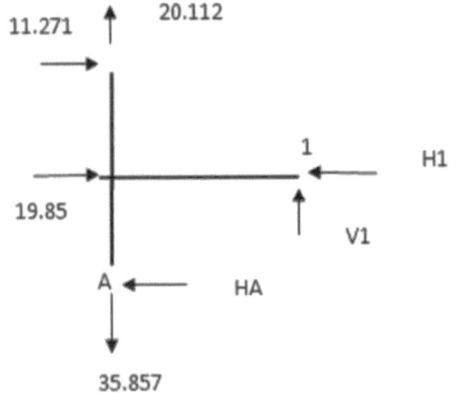

$\Sigma V = 0$
V1 = 15.745 kN
Taking the moment about 1:
HA = 14.684 kN
$\Sigma H = 0$
H1 = 16.437 kN.

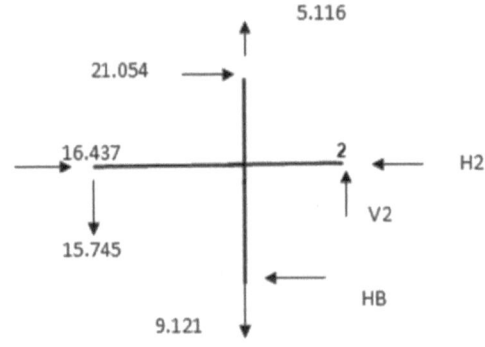

$\Sigma V = 0$
V2 = 19.75 kN
Taking the moment about 2:
HB = 27.428 kN
$\Sigma H = 0$
H2 = 10.063 kN.

(*continued*)

TABLE 4.40 (Continued)
Cantilever method of analysis

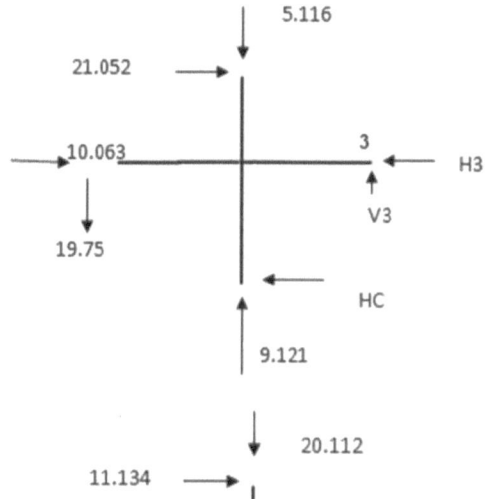

$\Sigma V = 0$
V3 = 15.745 kN
Taking the moment about 3:
Hc = 27.43 kN
$\Sigma H = 0$
H3 = 3.685 kN.

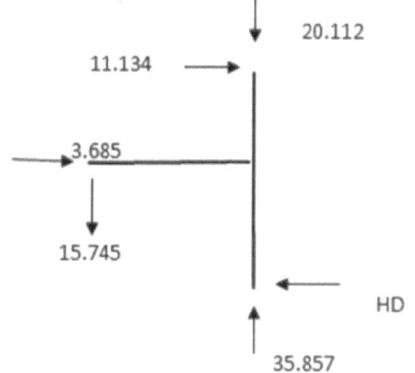

$\Sigma H = 0$
HD = 14.819 kN.

FBD of second floor

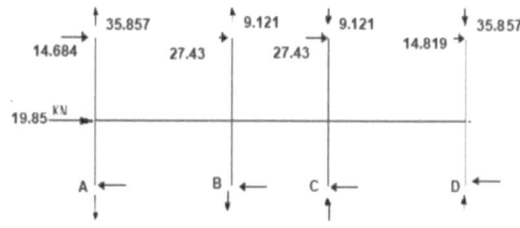

$\Sigma M = 0$
Taking the moment about D
19.85 x 1.6 + (14.684 + 27.43
+ 27.43 + 14.819)*3.2
+ 35.857*14.2 + 9.121*8.925
– 9.121*5.275 – $\Delta2$ x A x 14.2
– 0.25436 $\Delta2$ x A x (5.275 + 3.65)
+ 0.25436 x $\Delta2$ x 5.275 = 0 $\Delta2$
xA = 55.801 kN at A and D
Column load = 14.194 kN at B
and C

TABLE 4.40 (Continued)
Cantilever method of analysis

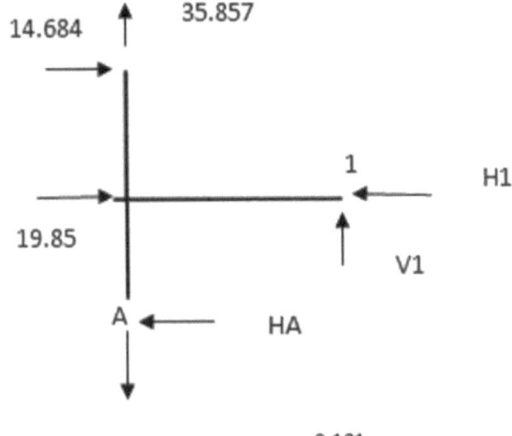

$\Sigma V = 0$
V1 =   19.944   kN
Taking the moment about 1:
HA =   18.192   kN
$\Sigma H = 0$
H1 =   16.342   kN

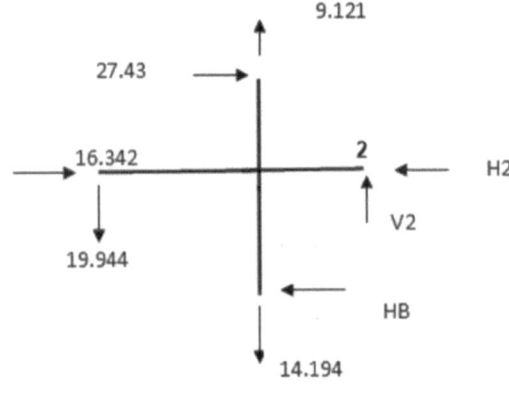

$\Sigma V = 0$
V2 =   25017   kN
Taking the moment about 2:
HB =   33.981   kN
$\Sigma H = 0$
H2 =   9.7905   kN

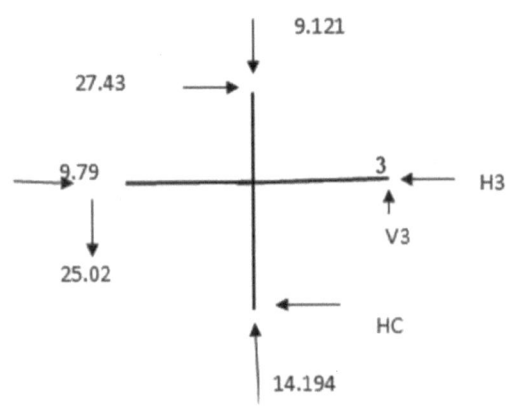

$\Sigma V = 0$
V3 = 19.947   kN
Taking the moment about 3:
Hc =   33.99   kN
$\Sigma H = 0$
H3 = 3.2302 kN

(continued)

TABLE 4.40 (Continued)
Cantilever method of analysis

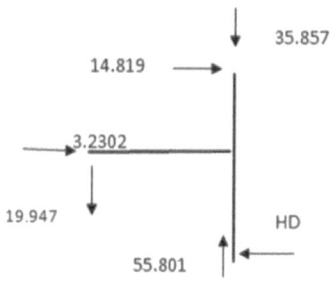

$\Sigma H = 0$
$HD = \quad 18.05 \quad kN$

FBD of first floor

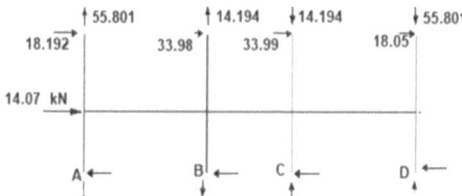

$\Sigma M = 0$
Taking moment about D
14.07 x 1.6 + (18.192 + 33.98
+ 33.99 + 18.05)*3.2 + 55.801*
14.2 + 14.194*8.925 – 14.194*
5.275 – Δ2 x A x 14.2 – 0.25436
Δ2 x A x (5.275 + 3.65) + 0.25436
x Δ2 x 5.275 = 0 Δ2 x A = 79.332
kN at A and D Column load =
20.179 kN at B and C

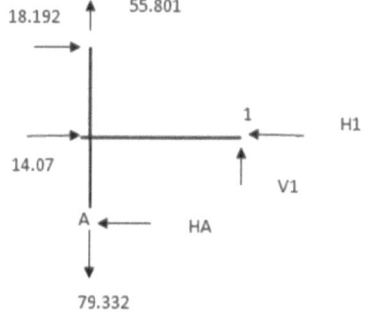

$\Sigma V = 0$
V1 = 23.531 kN
Taking the moment about 1:
HA = 20.597 kN
$\Sigma H = 0$
H1 = 11.665 kN

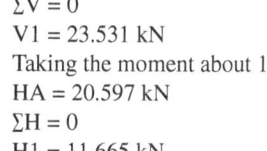

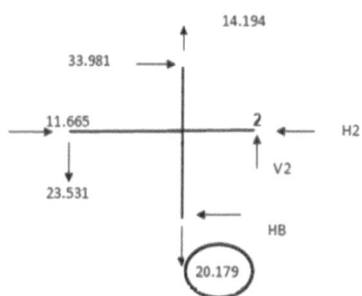

$\Sigma V = 0$
V2 = 29.516 kN
Taking the moment about 2:
HB = 38.475 kN
$\Sigma H = 0$
H2 = 7.1709 kN

TABLE 4.40 (Continued)
Cantilever method of analysis

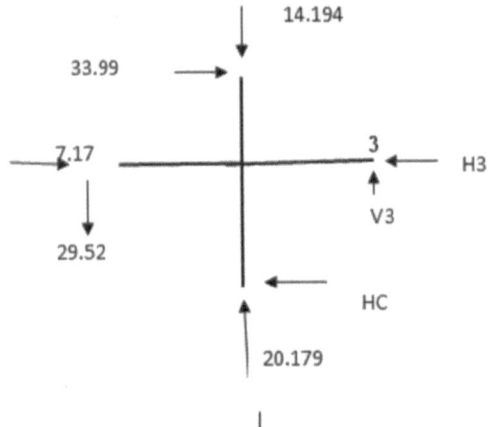

$\Sigma V = 0$
$V3 = 23.535$ kN
Taking the moment about 3:
$Hc = 38.477$ kN
$\Sigma H = 0$
$H3 = 2.6828$ kN

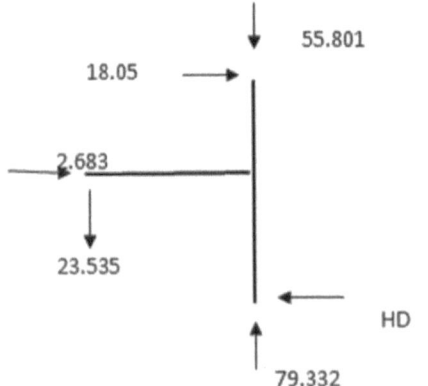

$\Sigma H = 0$
$HD = 20.733$ kN

FBD at plinth level

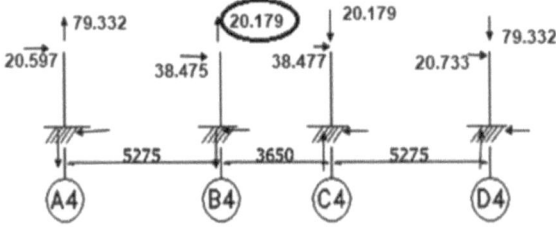

- **Encircled values are referred in the calculations made later on.**

The design bending moments and shear force values in the shorter frame (4–4) are obtained and furnished in Tables 4.41, 4.42 and 4.43, referring to the calculations made in Table 4.40.

TABLE 4.41
Design column moment and shear force

	Column A4		Column B4		Column C4		Column D4	
Level	Shear (kN)	Moment (kNm)	Shear (kN)	Moment (kNm)	Shear (kN)	Moment (kNm)	Shear (kN)	Moment (kNm)
Ground level	20.60	37.10	38.50	69.30	38.50	69.30	20.70	37.30
First floor lower	−20.60	37.10	−38.50	69.30	−38.50	69.30	−20.70	37.30
First floor upper	18.20	29.10	33.90	61.02	33.90	61.02	18.10	29.00
Second floor lower	−18.20	29.10	−33.90	61.02	−33.90	61.02	−18.10	29.00
Second floor upper	14.70	23.50	27.43	43.90	27.43	43.90	14.80	23.70
Third floor lower	−14.70	23.50	−27.43	43.90	−27.43	43.90	−14.80	23.70
Third floor upper	11.27	17.98	21.05	33.70	21.05	33.70	11.13	17.80
Fourth floor lower	−11.27	17.98	−21.05	33.70	−21.05	33.70	−11.13	17.80
Fourth floor upper	7.48	11.97	13.97	22.40	13.97	22.40	7.55	12.10
Fifth floor lower	−7.48	11.97	−13.97	22.40	−13.97	22.40	−7.55	12.10
Fifth floor upper	3.46	5.54	6.47	10.40	6.47	10.40	3.49	5.54
Sixth floor lower	−3.46	5.54	−6.47	10.40	−6.47	10.40	−3.49	5.54

TABLE 4.42
Design beam moment and shear force

First floor level	Shear (kN)	Moment (kNm)
Beam A4–B4		
Left	−23.5	−62
Right	23.5	−62
Beam B4–C4		
Left	−29.5	−54
Right	29.5	−54
Beam C4–D4		
Left	−23.5	−62
Right	23.5	−62

TABLE 4.43
Axial force and bending moment of ground floor column under wind load in kN (refer to Table 4.40 at plinth level)

Column	A4	B4	C4	D4
Axial force in kN	±79.3	±20.2	±20.2	±79.3
Moment in kNm	±37.1	±69.3	±69.3	±37.3

Notes: Encircled values are referred to in the calculations made later. The ± sign indicates that winds from both left and right are considered. Frame analysis in the longer direction under wind loading and frame analysis under seismic loading in both the longer and shorter directions may be carried out in the same manner; however, the remaining frame analyses have been made using software.

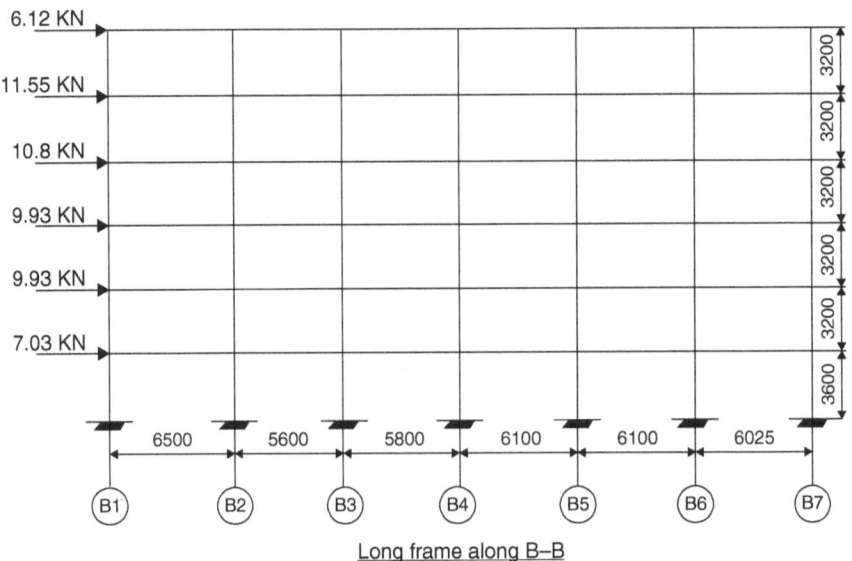

FIGURE 4.35 Frame in the longer direction (B–B) showing wind loads

The wind loads at different levels have been calculated for the longer frame (B–B), and are shown in Figure 4.35.

The design beam moments and shear forces in the longer frame (B–B) have been calculated and furnished in Tables 4.44, 4.45 and 4.46.

TABLE 4.44
Design column moment and shear force

Level	Column B1		Column B2		Column B3		Column B4		Column B5		Column B6		Column B7	
	Shear (kN)	Moment (kNm)	Shear (kN)	Moment (kNm)	Shear (kN)	Moment (kNm)	Shear (kN)	Moment (kNm)	Shear (kN)	Moment (kNm)	Shear (kN)	Moment (kNm)	Shear (kN)	Moment (kNm)
Ground level	6.069	10.924	9.484	17.072	7.900	14.219	8.881	15.986	7.697	13.854	9.314	16.766	6.015	10.826
First floor lower	-6.069	10.924	-9.484	17.072	-7.900	14.219	-8.881	15.986	-7.697	13.854	-9.314	16.766	-6.015	10.826
First floor upper	4.305	6.888	8.351	13.362	7.834	12.534	7.527	12.043	7.544	12.071	8.287	13.259	4.482	7.171
Second floor lower	-4.305	6.888	-8.351	13.362	-7.834	12.534	-7.527	12.043	-7.544	12.071	-8.287	13.259	-4.482	7.171
Second floor upper	3.629	5.806	6.505	10.408	5.992	9.587	6.163	9.861	5.823	9.317	6.532	10.451	3.756	6.009
Third floor lower	-3.629	5.806	-6.505	10.408	-5.992	9.587	-6.163	9.861	-5.823	9.317	-6.532	10.451	-3.756	6.009
Third floor upper	2.617	4.187	4.874	7.799	4.527	7.244	4.512	7.220	4.403	7.044	4.848	7.757	2.688	4.301
Fourth floor lower	-2.617	4.187	-4.874	7.799	-4.527	7.244	-4.512	7.220	-4.403	7.044	-4.848	7.757	-2.688	4.301
Fourth floor upper	1.650	2.640	3.014	4.823	2.803	4.485	2.837	4.539	2.737	4.380	2.984	4.775	1.644	2.631
Fifth floor lower	-1.650	2.640	-3.014	4.823	-2.803	4.485	-2.837	4.539	-2.737	4.380	-2.984	4.775	-1.644	2.631
Fifth floor upper	0.386	0.618	0.936	1.498	1.046	1.674	1.011	1.618	1.112	1.779	1.113	1.780	0.516	0.825
Sixth floor lower	-0.386	0.618	-0.936	1.498	-1.046	1.674	-1.011	1.618	-1.112	1.779	-1.113	1.780	-0.516	0.825

TABLE 4.45
Design beam moment and shear force

First floor level	Shear (kN)	Moment (kNm)
Beam B1–B2		
Support B1 (left)	–5.481	–17.812
Support B2 (left)	5.481	–17.812
Beam B2–B3		
Support B2 (right)	–4.508	–12.622
Support B3 (left)	4.508	–12.622
Beam B3–B4		
Support B3 (right)	–4.873	–14.132
Support B4 (left)	4.873	–14.132
Beam B4–B5		
Support B4 (right)	–4.557	–13.898
Support B5 (left)	4.557	–13.898
Beam B5–B6		
Support B5 (right)	–3.943	–12.027
Support B6 (left)	3.943	–12.027
Beam B6–B7		
Support B6 (right)	–5.974	–17.998
Support B7 (left)	5.974	–17.998

TABLE 4.46
Axial force and bending moment of ground floor column under wind load in kN

Column	B1	B2	B3	B4	B5	B6	B7
Axial force in kN	±15.80	±0.90	±0.20	⦵±0.97⦵	±0.92	±4.30	±17.5
Moment in kNm	±10.90	±17.10	±14.20	⦵±15.98⦵	±13.85	±16.67	±10.82

Note: Encircled values are referred to in the calculations made later.

The seismic loads at different levels have been calculated for the shorter frame (4–4), and are shown in Figure 4.36.

The design column moments and shear forces are shown in Tables 4.47, 4.48 and 4.49.

The seismic loads at different levels have been calculated for the longer frame (B–B), and are shown in Figure 4.37.

The design column moments and shear forces are shown in Tables 4.50, 4.51 and 4.52.

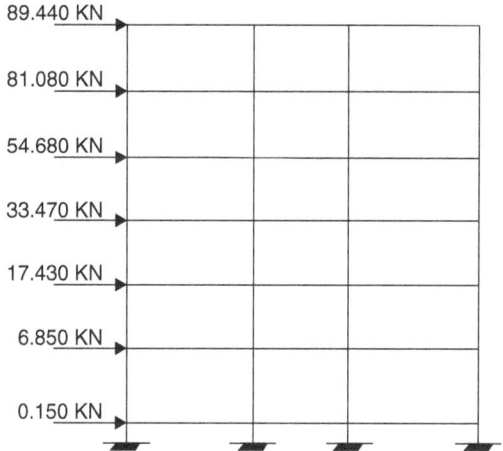

89.440 KN

81.080 KN

54.680 KN

33.470 KN

17.430 KN

6.850 KN

0.150 KN

FIGURE 4.36 Frame in the shorter direction showing seismic loads

4.4.11 SUMMARY ON MAXIMUM BENDING MOMENT AND SHEAR DUE TO DEAD LOAD, LIVE LOAD, WIND LOAD AND SEISMIC LOAD

Dead and Live Loads

The design values for shorter frames are shown in Tables 4.53, 4.54, 4.55 and 4.56. The maximum forces due to dead and live loads are shown in Tables 4.57, 4.58, 4.59 and 4.60.

The maximum axial force, bending moment and shear force in beams of shorter frame due to wind loads are shown in Tables 4.61 and 4.62.

The maximum axial force, bending moment and shear force in beams of the longer frame due to wind loads are shown in Tables 4.63 and 4.64.

The maximum axial force, bending moment and shear force in beams of the shorter frame due to seismic loads are shown in Tables 4.65 and 4.66.

The maximum axial force, bending moment and shear force in beams of the longer frame due to seismic loads are shown in Tables 4.67 and 4.68.

Let us now design the column and beam members using the limit state method.

Different load cases created by combining different loads with partial safety factors, as per Table 18 of IS 456, 2000, are furnished in Tables 4.69, 4.70, 4.71, 4.72 and 4.73.

TABLE 4.47
Design column moment and shear force

Level	Column A4		Column B4		Column C4		Column D4	
	Shear (kN)	Moment (kNm)	Shear (kN)	Moment (kNm)	Shear (kN)	Moment (kNm)	Shear (kN)	Moment (kNm)
Foundations	59.435	195.865	82.093	270.906	82.055	270.782	59.247	195.516
Tie beam lower	−59.435	−106.713	−82.093	−147.767	−82.055	−147.699	−59.247	−106.645
Tie beam upper	59.285	106.713	82.093	147.767	82.055	147.699	59.247	106.645
First floor lower	−59.285	106.713	−82.093	147.767	−82.055	147.699	−59.247	106.645
First floor upper	32.486	51.977	105.665	169.064	105.564	168.903	32.385	51.816
Second floor lower	−32.486	51.977	−105.665	169.064	−105.564	168.903	−32.385	51.817
Second floor upper	46.905	75.048	82.604	132.167	82.43	131.888	46.731	74.769
Third floor lower	−46.905	75.048	−82.604	132.167	−82.43	131.888	−46.731	74.769
Third floor upper	32.404	51.846	80.346	128.554	80.196	128.314	32.254	51.606
Fourth floor lower	−32.404	51.846	−80.346	128.554	−80.196	128.314	−32.254	51.606
Fourth floor upper	29.213	46.74	56.443	90.309	56.047	89.676	28.817	46.107
Fifth floor lower	−29.213	46.74	−56.443	90.309	−56.047	89.676	−28.817	46.107
Fifth floor upper	14.988	23.981	29.705	47.528	29.732	47.571	15.015	24.024
Sixth floor lower	−14.988	23.981	−29.705	47.528	−29.732	47.571	−15.015	24.024

TABLE 4.48
Design beam moment and shear

First floor level	Shear (kN)	Moment (kNm)
Beam A4–B4		
Support A (left)	–60.2	–158.7
Support B (left)	60.2	–158.7
Beam B4–C4		
Support B (right)	–86.7	–158.1
Support C (left)	86.7	–158.1
Beam C4–D4		
Support C (right)	–60.1	–158.5
Support D (left)	60.1	–158.5

TABLE 4.49
Axial force and bending moment of ground floor column under wind load in kN

Column	A4	B4	C4	D4
Axial force in kN	±229.7	±141.3	±142.3	±228.7
Moment in kNm	±195.9	±270.9	±270.8	±195.5

Note: Encircled values are referred to in the calculations made later.

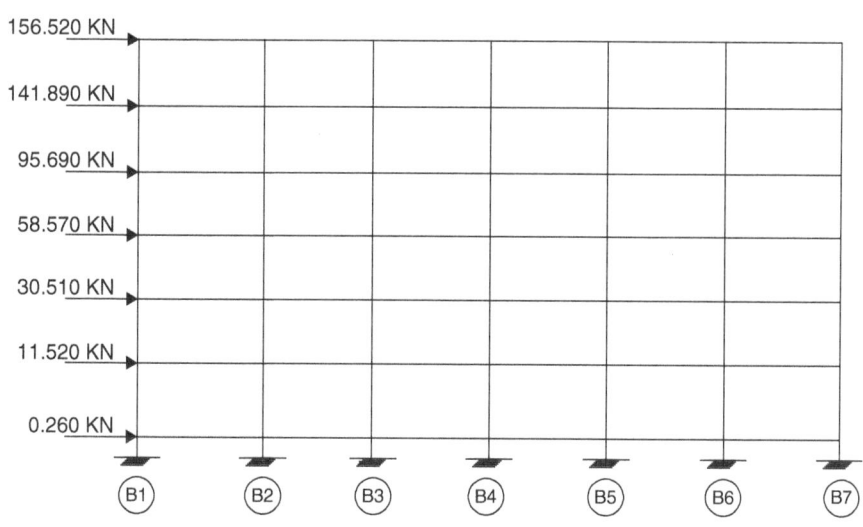

FIGURE 4.37   Frame in the longer direction showing seismic loads

TABLE 4.50
Design column moment and shear force

Lev	Column B1		Column B2		Column B3		Column B4		Column B5		Column B6		Column B7	
	Shear (kN)	Moment (kNm)	Shear (kN)	Moment (kNm)	Shear (kN)	Moment (kNm)	Shear (kN)	Moment (kNm)	Shear (kN)	Moment (kNm)	Shear (kN)	Moment (kNm)	Shear (kN)	Moment (kNm)
Foundation	58.946	194.052	83.066	274.119	64.993	214.477	81.23	268.059	64.597	213.17	83.054	274.077	59.074	194.946
Ground level lower	-58.946	-105.634	-83.066	-149.519	-64.993	-116.988	-81.23	-146.214	-64.597	-116.274	-83.054	-149.497	-59.074	-106.334
Ground level upper	58.686	105.634	83.066	149.519	64.993	116.988	81.23	146.214	64.597	116.274	83.054	149.497	59.074	106.334
First Floor lower	-58.686	105.634	-83.066	149.519	-64.993	116.988	-81.23	146.214	-64.597	116.274	-83.054	149.497	-59.074	106.334
First Floor upper	41.331	66.13	83.751	134.002	79.957	127.932	74.349	118.958	77.155	123.448	83.49	133.584	43.146	69.034
Second Floor lower	-41.331	66.13	-83.751	134.002	-79.957	127.932	-74.349	118.958	-77.155	123.448	-83.49	133.584	-43.146	69.034
Second Floor upper	43.354	69.367	77.083	123.332	70.546	112.873	72.902	116.644	68.358	109.373	76.35	122.16	44.077	70.524
Third Floor lower	-43.354	69.367	-77.083	123.332	-70.546	112.873	-72.902	116.644	-68.358	109.373	-76.35	122.16	-44.077	70.524
Third Floor upper	36.462	58.34	67.865	108.585	62.898	100.637	62.369	99.79	60.877	97.403	66.816	106.905	36.813	58.901
Fourth Floor lower	-36.462	58.34	-67.865	108.585	-62.898	100.637	-62.369	99.79	-60.877	97.403	-66.816	106.905	-36.813	58.901
Fourth Floor upper	28.154	45.047	51.491	82.386	47.47	75.952	47.715	76.345	45.881	73.409	49.998	79.997	27.7	44.32
Fifth Floor lower	-28.154	45.047	-51.491	82.386	-47.47	75.952	-47.715	76.345	-45.881	73.409	-49.998	79.997	-27.7	44.32
Fifth Floor upper	13.711	21.938	26.826	42.921	25.813	41.301	25.084	40.134	25.397	40.636	26.351	42.161	13.338	21.341
Sixth Floor lower	-13.711	21.938	-26.826	42.921	-25.813	41.301	-25.084	40.134	-25.397	40.636	-26.351	42.161	-13.338	21.341

TABLE 4.51
Design beam moment and shear force

First floor level	Shear (kN)	Moment (kNm)
Beam B1–B2		
Support B1 (left)	−53.0	−171.8
Support B2 (left)	53.0	−171.8
Beam B2–B3		
Support B2 (right)	−40.0	−111.8
Support B3 (left)	40.0	−111.8
Beam B3–B4		
Support B3 (right)	−46.0	−133.2
Support B4 (left)	46.0	−133.2
Beam B4–B5		
Support B4 (right)	−43.3	−132.0
Support B5 (left)	43.3	−132.0
Beam B5–B6		
Support B5 (right)	−35.3	−107.7
Support B6 (left)	35.3	−107.7
Beam B6–B7		
Support B6 (right)	−58.2	−175.4
Support B7 (left)	58.2	−175.4

TABLE 4.52
Axial force and bending moment of ground floor column under wind load in kN

Column	B1	B2	B3	B4	B5	B6	B7
Axial force in kN	±193.0	±12.8	±2.4	±12.1	±11.72	±51.8	±210.6
Moment in kNm	±194.0	±274	±214.5	±268.0	±213.2	±274.0	±195.0

Note: Encircled values are referred to in the calculations made later.

TABLE 4.53
Bending moment due to dead and live loads (kNm) (same as Table 4.29)

Support	A		B_{left}	B_{right}		C_{left}	C_{right}		D
Span		AB			BC			CD	
Support moment	−67.50		−108.70	−62.30		−62.30	−108.70		−67.50
Span moment		64.67			17.80			64.67	

TABLE 4.54
Shear force due to dead and live loads (kN) (same as Table 4.30)

	A		B		C		D	
Support	Left	Right	Left	Right	Left	Right	Left	Right
Shear force		102.1	−109.2	54.8	−54.8	109.2	−102.1	

Note: Encircled values are referred to in the calculations made later.

TABLE 4.55
Bending moment due to dead load only (kNm) (same as Table 4.27)

Support	A		B_{left}	B_{right}		C_{left}	C_{right}		D
Span		AB			BC			CD	
Support moment	−71.3		−74.1	−27.9		−27.9	−74.1		−71.3
Span moment		45.1			11.2			45.1	

TABLE 4.56
Shear force due to dead load only (kN)(same as Table 4.28)

Support	A		B		C		D	
	Left	Right	Left	Right	Left	Right	Left	Right
Shear force		73.2	−74.2	36.8	−36.8	74.2	−73.2	

Note: Encircled values are referred to in the calculations made later.

TABLE 4.57
Bending moment due to dead and live loads (kNm) (same as Table 4.38)

	1		2		3		4	
Support	Right		Left	Right		Left	Right		Left	Right
Span		1–2			2–3			3–4			4–5
Support moment	−159.7		−189.0	−119.4		−120.4	−127.9		−139.1	−146.8
Span moment		160			82.2			79.4			89.3
Support		5			6			7	
	Left	Right		Left	Right		Left	
Span		5–6			6–7			
Support moment	−153.8	−157.9		−151.9	−134.8		−70.3	
Span moment		108.8			82.8			

TABLE 4.58
Shear force due to dead and live loads (kN) (same as Table 4.39)

	1		2		3		4		5		6		7
Support	Left	Right	Left	Right	Left	Right	Left	Right	Left	Right	Left	Right	Left
Shear force	130.2	−163.8	117.8	−106.3	114.2	(−118.1)	(124.0)	−125.9	128.3	−127.0	106.6	−85.2	

Note: Encircled values are referred to in the calculations made later.

TABLE 4.59
Bending moment due to dead load only (kNm) (same as Table 4.36)

	1		2		3		4	
Support	Right		Left	Right		Left	Right		Left	Right
Span	1–2		2–3		3–4		4–5	
Support moment	−105.0		−152.9	−107.0		−77.2	−85.6		−95.8	−100.6
Span moment		120.5		48.2		51		57.6
Support	5		6		7			
	Left	Right		Left	Right		Left	
Span		5–6		6–7			
Support moment	−105.2	−106.8		−106.8	−105.2		−59.8	
Span moment		69.3		64.2			

TABLE 4.60
Shear force due to dead load only (kN) (same as Table 4.37)

	1		2		3		4		5		6		7
Support	Left	Right	Left	Right	Left	Right	Left	Right	Left	Right	Left	Right	Left
Shear force	104.1	−126.2	81.9	−71.2	80.3	(−84.0)	(87.7)	−89.1	89.4	−89.4	87.9	−72.8	

Note: Encircled values are referred to in the calculations made later.

TABLE 4.61
Bending moment and shear force of first floor beams due to wind load (kN) (same as Table 4.42)

First floor level	Shear(kN)	Moment (kNm)
Beam A4–B4		
Left	−23.5	−62
Right	23.5	−62
Beam B4–C4		
Left	−29.5	−54
Right	29.5	−54
Beam C4–D4		
Left	−23.5	−62
Right	23.5	−62

TABLE 4.62
Axial force of ground floor column due to wind load (kN) (same as Table 4.43)

Column	A4	B4	C4	D4
Axial force in kN	±79.3	±20.2	±20.2	±79.3
Moment in kNm	±37.1	±69.3	±69.3	±37.3

Note: Encircled values are referred to in the calculations made later.

TABLE 4.63
Bending moment and shear force of first floor beams due to wind load (kN) (same as Table 4.45)

First floor level	Shear (kN)	Moment (kNm)
Beam B1–B2		
Support B1 (left)	−5.481	−17.812
Support B2 (left)	5.481	−17.812
Beam B2–B3		
Support B2 (right)	−4.508	−12.622
Support B3 (left)	4.508	−12.622
Beam B3–B4		
Support B3 (right)	−4.873	−14.132
Support B4 (left)	4.873	−14.132

(continued)

TABLE 4.63 (Continued)
Bending moment and shear force of first floor beams due to
wind load (kN) (same as Table 4.45)

First floor level	Shear (kN)	Moment (kNm)
Beam B4–B5		
Support B4 (right)	–4.557	–13.898
Support B5 (left)	4.557	–13.898
Beam B5–B6		
Support B5 (right)	–3.943	–12.027
Support B6 (left)	3.943	–12.027
Beam B6–B7		
Support B6 (right)	–5.974	–17.998
Support B7 (left)	5.974	–17.998

TABLE 4.64
Axial force of ground floor column for wind load (kN) (same as Table 4.46)

Column	B1	B2	B3	B4	B5	B6	B7
Axial force in kN	±15.80	±0.90	±0.20	(±0.97)	±0.92	±4.30	±17.50
Moment in kNm	±10.90	±17.10	±14.20	(±15.98)	±13.85	±16.67	±10.82

Note: Encircled values are referred to in the calculations made later.

TABLE 4.65
Bending moment and shear force of first floor beams for seismic
load (kNm) (same as Table 4.48)

First floor level	Shear(kN)	Moment (kNm)
Beam A4–B4		
Support A (left)	–60.2	–158.7
Support B (left)	60.2	–158.7
Beam B4–C4		
Support B (right)	–86.7	–158.1
Support C (left)	86.7	–158.1
Beam C4–D4		
Support C (right)	–60.1	–158.5
Support D (left)	60.1	–158.5

TABLE 4.66
Axial force of ground floor column for seismic load (kN) (same as Table 4.49)

Column	A4	B4	C4	D4
Axial force in kN	±229.7	±141.3	±142.3	±228.7
Moment in kNm	±195.9	±270.9	±270.8	±195.5

Note: Encircled values are referred to in the calculations made later.

TABLE 4.67
Bending moment and shear force of first floor beams due to seismic load (kN) (same as Table 4.51)

First floor level	Shear (kN)	Moment (kNm)
Beam B1–B2		
Support B1 (left)	–53.0	–171.8
Support B2 (left)	53.0	–171.8
Beam B2–B3		
Support B2 (right)	–40.0	–111.8
Support B3 (left)	40.0	–111.8
Beam B3–B4		
Support B3 (right)	–46.0	–133.2
Support B4 (left)	46.0	–133.2
Beam B4–B5		
Support B4 (right)	–43.3	–132.0
Support B5 (left)	43.3	–132.0
Beam B5–B6		
Support B5 (right)	–35.3	–107.7
Support B6 (left)	35.3	–107.7
Beam B6–B7		
Support B6 (right)s	–58.2	–175.4
Support B7 (left)	58.2	–175.4

TABLE 4.68
Axial force of ground floor column for seismic load (kN) (same as Table 4.52)

Column	B1	B2	B3	B4	B5	B6	B7
Axial force in kN	±193.00	±12.80	±2.40	±12.10	±11.72	±51.80	±210.60
Moment in kNm	±194.00	±274.00	±214.5	±268.00	±213.20	±274.00	±195.00

Note: Encircled values are referred to in the calculations made later.

TABLE 4.69
Load combination

	Load factors					
Loadcase	DL	LL	WL-X	WL-Y	EQ-X	EQ-Y
Loadcase 1	1.5	1.5				
Loadcase 2	1.5		1.5			
Load case 3	1.5		−1.5			
Load case 4	1.5			1.5		
Load case 5	1.5			−1.5		
Load case 6	1.5				1.5	
Load case 7	1.5				−1.5	
Load case 8	1.5					1.5
Load case 9	1.5					−1.5
Load case 10	1.2	1.2	1.2			
Load case 11	1.2	1.2	−1.2			
Load case 12	1.2	1.2		1.2		
Load case 13	1.2	1.2		−1.2		
Load case 14	1.2	1.2			1.2	
Load case 15	1.2	1.2			−1.2	
Load case 16	1.2	1.2				1.2
Load case 17	1.2	1.2				−1.2
Load case 18	0.9				1.5	
Load case 19	0.9				−1.5	
Load case 20	0.9					1.5
Load case 21	0.9					−1.5

Notes: Refer to Tables 4.29/4.53, 4.30/4.54, 4.27/4.55, 4.28/4.56, 4.42/4.61, 4.43/4.62, 4.48/4.65 and 4.49/4.66. WL-X = the wind force acting along the longer direction on the shorter face. WL-Y = the wind force acting along the shorter direction on the longer face. EQ-X = the seismic force acting along the longer direction on the shorter face. EQ-Y = the seismic force acting along the longer direction on the shorter face.

TABLE 4.70
Load combination of bending moments for the shorter frame beam (4–4) (first floor)

Load case	A	B		C		D
		Left of support B	Right of support B	Left of support C	Right of support C	
1	153.15	−163.80	82.20	−82.20	163.80	−153.15
2		Refer to the longer frame analysis (B–B)				
3		Refer to the longer frame analysis (B–B)				
4	74.55	−146.55	10.95	−99.45	76.05	−145.05
5	145.05	−76.05	99.45	−10.95	146.55	−74.55
6		Refer to the longer frame analysis (B–B)				
7		Refer to the longer frame analysis (B–B)				

(continued)

TABLE 4.70 (Continued)
Load combination

Load case	A	B		C		D
		Left of support B	Right of support B	Left of support C	Right of support C	
8	19.50	*−201.6*	−74.85	*−185.25*	21.15	*−199.95*
9	*200.10*	−21.00	*185.25*	74.85	*201.45*	−19.65
10		Refer to the longer frame analysis (B–B)				
11		Refer to the longer frame analysis (B–B)				
12	94.32	−159.24	30.36	−101.16	102.84	−150.72
13	150.72	−102.84	101.16	−30.36	159.24	−94.32
14		Refer to the longer frame analysis (B–B)				
15		Refer to the longer frame analysis (B–B)				
16	50.28	−203.28	−38.28	−169.80	58.92	−194.64
17	194.76	−58.8	169.8	38.28	203.16	−50.40
18		Refer to the longer frame analysis (B–B)				
19		Refer to the longer frame analysis (B–B)				
20	−24.42	−157.08	−53.58	−119.82	−23.37	−156.03
21	156.18	23.52	119.82	53.58	156.93	24.27

Note: Maximum values are denoted with italic script.

TABLE 4.71
Load combination of shear forces for the shorter frame beam (4–4) (first floor)

Load case	AB			BC			CD		
	Left support	Span	Right support	Left support	Span	Right support	Left support	Span	Right support
1	−101.250	97.005	−163.050	−93.450	26.700	−93.450	−163.050	97.005	−101.250
2				Refer to the longer frame analysis (B–B)					
3				Refer to the longer frame analysis (B–B)					
4	−13.950	67.650	−204.150	39.150	16.800	−122.850	−18.150	67.650	−199.950
5	−199.950	67.650	−18.150	−122.850	16.800	39.150	−204.150	67.650	−13.950
6				Refer to the longer frame analysis (B–B)					
7				Refer to the longer frame analysis (B–B)					
8	131.100	67.650	−349.200	195.300	16.800	−279.000	126.600	67.650	−344.700
9	−345.000	67.650	126.900	−279.000	16.800	195.300	−348.900	67.650	130.800
10				Refer to the longer frame analysis (B–B)					
11				Refer to the longer frame analysis (B–B)					
12	−6.600	77.604	−204.840	−9.960	21.360	−139.560	−56.040	77.604	−155.400
13	−155.400	77.604	−56.040	−139.560	21.360	−9.960	−204.840	77.604	−6.600
14				Refer to the longer frame analysis (B–B)					
15				Refer to the longer frame analysis (B–B)					
16	109.440	77.604	−320.880	114.960	21.360	−264.480	59.760	77.604	−271.200
17	271.440	77.604	60.000	−264.480	21.360	114.960	−320.640	77.604	109.200
18				Refer to the longer frame analysis (B–B)					
19				Refer to the longer frame analysis (B–B)					
20	177.300	58.203	−335.880	181.080	16.020	−214.170	139.920	58.203	−298.500
21	−298.800	58.203	140.220	−293.220	16.020	102.030	−335.580	58.203	177.00

Note: Maximum values are denoted with italic script.

TABLE 4.72
Load combination of bending moments for the longer frame beam (first floor)

Load case	1–2 Left of support	1–2 Span	1–2 Right of support	2–3 Left of support	2–3 Span	2–3 Right of support	3–4 Left of support	3–4 Span	3–4 Right of support	4–5 Left of support	4–5 Span	4–5 Right of support	5–6 Left of support	5–6 Span	5–6 Right of support	6–7 Left of support	6–7 Span	6–7 Right of support
1	−254.85	*240.00*	−282.15	−168.45	*123.30*	−180.60	*−191.85*	*119.10*	−208.65	−220.20	133.95	−230.70	−236.85	163.20	−227.85	−202.20	124.20	−105.45
2	−184.20	180.75	−202.65	−179.40	72.30	−96.90	−149.55	76.50	−122.55	−171.75	86.40	−136.95	−178.20	103.95	−142.20	−184.80	96.30	−62.70
3	−130.80	180.75	−256.05	−141.60	72.30	−134.70	−107.25	76.50	−164.85	−130.05	86.40	−178.65	−142.20	103.95	−178.20	−130.80	96.30	−116.70
4							Refer to the shorter frame analysis (4–4)											
5							Refer to the shorter frame analysis (4–4)											
6	100.20	180.75	−487.05	7.20	72.30	−283.50	71.40	76.50	−343.50	47.10	86.40	−355.80	1.35	103.95	−321.75	105.30	96.30	−352.80
7	−415.20	180.75	28.35	−328.20	72.30	51.90	−328.20	76.50	56.10	−348.90	86.40	40.20	−321.75	103.95	1.35	−420.90	96.30	173.40
8							Refer to the shorter frame analysis (4–4)											
9							Refer to the shorter frame analysis (4–4)											
10	−225.24	192.00	−204.36	−149.88	98.64	−129.36	−170.40	95.28	−150.00	−192.84	107.16	−167.88	−203.88	130.56	−167.88	−183.36	99.36	−62.76
11	−182.52	192.00	−247.08	−119.64	98.64	−159.60	−136.56	95.28	−183.84	−159.48	107.16	−201.24	−175.08	130.56	−196.68	−140.16	99.36	−105.96
12							Refer to the shorter frame analysis (4–4)											
13							Refer to the shorter frame analysis (4–4)											
14	2.28	192.00	−431.88	−0.60	98.64	−278.64	6.36	95.28	−326.76	−17.76	107.16	−342.96	−60.24	130.56	−311.52	48.72	99.36	−294.84
15	−410.04	192.00	−19.56	−268.92	98.64	−10.32	−313.32	95.28	−7.08	−334.56	107.16	−26.16	−318.72	130.56	−53.04	−372.24	99.36	126.12
16							Refer to the shorter frame analysis (4–4)											
17							Refer to the shorter frame analysis (4–4)											
18	163.20	108.45	−395.31	71.40	43.38	−237.18	122.76	45.90	−286.02	107.46	51.84	−292.68	65.43	62.37	−257.67	168.42	57.78	−316.92
19	−352.20	108.45	120.09	−264.00	43.38	98.22	−276.84	45.90	113.58	−288.54	51.84	103.32	−257.67	62.37	65.43	−357.78	57.78	209.28
20							Refer to the shorter frame analysis (4–4)											
21							Refer to the shorter frame analysis (4–4)											

Notes: Refer to Tables 4.38/4.57, 4.39/4.58, 4.36/4.59, 4.37/4.60, 4.45/4.63, 4.46/4.64, 4.51/4.67 and 4.52/4.68. Maximum values are denoted with italic script.

TABLE 4.73

Load combination of shear forces for the longer frame beam

Load case	1	2		3		4		5		6		7
		Left of support	Right of support	Left of support	Right of support	Left of support	Right of support	Left of support	Right of support	Left of support	Right of support	
1	195.30	−245.70	176.70	−159.45	171.30	−177.15	186.00	−188.85	192.45	−190.50	159.90	−127.80
2	147.90	−197.55	116.10	−113.55	113.10	−133.35	124.65	−140.55	128.25	−139.95	122.90	−118.16
3	164.40	−181.05	129.60	−100.05	127.80	−118.65	138.45	−126.75	139.95	−128.25	140.81	−100.25
4					Refer to the shorter frame analysis (4-4)	Refer to the shorter frame analysis (4-4)						
5					Refer to the shorter frame analysis (4-4)	Refer to the shorter frame analysis (4-4)						
6	*76.65*	*−268.80*	*62.85*	*−166.80*	*51.45*	*−195.00*	*66.60*	*−198.60*	*81.15*	*−187.10*	*44.55*	*−196.50*
7	*235.7*	*−109.80*	*182.85*	*−46.80*	*189.45*	*−57.00*	*196.50*	*−68.70*	*187.05*	*−81.15*	*219.20*	*−21.90*
8					Refer to the shorter frame analysis (4-4)	Refer to the shorter frame analysis (4-4)						
9					Refer to the shorter frame analysis (4-4)	Refer to the shorter frame analysis (4-4)						
10	149.64	−203.16	135.96	−132.96	131.16	−147.60	143.28	−156.60	149.28	−157.08	120.76	−109.40
11	135.70	−158.30	122.30	−101.80	119.10	−113.20	128.60	−121.30	132.20	−123.10	112.57	−79.23
12					Refer to the shorter frame analysis (4-4)	Refer to the shorter frame analysis (4-4)						
13					Refer to the shorter frame analysis (4-4)	Refer to the shorter frame analysis (4-4)						
14	92.64	−260.16	93.36	−175.56	81.84	−196.92	96.84	−203.04	111.60	−194.76	58.08	−172.08
15	219.84	−132.96	189.36	−79.56	192.24	−86.52	200.76	−99.12	196.32	−110.04	197.76	−32.40
16					Refer to the shorter frame analysis (4-4)	Refer to the shorter frame analysis (4-4)						
17					Refer to the shorter frame analysis (4-4)	Refer to the shorter frame analysis (4-4)						
18	14.19	−193.08	13.71	−124.08	3.27	−144.60	13.98	−145.14	27.51	−133.41	−8.19	−152.82
19	173.19	−34.08	133.71	−4.08	141.27	−6.60	143.88	−15.24	133.41	−27.51	166.41	21.78
20					Refer to the shorter frame analysis (4-4)	Refer to the shorter frame analysis (4-4)						
21					Refer to the shorter frame analysis (4-4)	Refer to the shorter frame analysis (4-4)						

Note: Maximum values are denoted with italic script.

4.4.12 Design of Frame Beams

The effective depth of beam is to be calculated as per IS clause 23 of IS 456, 2000: "Effective depth of a beam is the distance between the centroid of the area of tension reinforcement and the maximum compression fiber, excluding the thickness of finishing material not placed monolithically with the member and the thickness of any concrete provided to allow for wear."

So, effective depth = d = total depth (D) – nominal cover –diameter of stirrups – half diameter of main bar.

As per table 16 of IS 456, 2000, clause 26.4.2: the nominal cover to meet durability requirements, for moderate exposure, is 30 mm.

As per table 16A of IS 456, 2000, clauses 26.4.3 and 21.4: the nominal cover to meet the specified fire resistance, for a continuous beam with two hours' fire resistance, is 30 mm.

The value of M_u/bd^2 has been calculated and checked with M_{ulim}/bd^2 for determining whether the beam is a singly or a doubly reinforced section. From SP 16, table 2C, the value can be calculated.

For $f_y = 500$ N/mm^2 and $f_{ck} = M_{ulim}/bd^2 = 0.133\ f_{ck} = 0.133$ X 25 = 3.325 N/mm^2

For a single reinforced beam with $f_{ck} = 25$ N/mm^2 and $f_y = 500$ N/mm^2, the reinforcement percentage can be obtained using Table 3 of SP 16.

For a double reinforced beam with $f_{ck} = 25$ N/mm^2 and $f_y = 500$ N/mm^2, the reinforcement percentage can be obtained using Table 55 of SP 16.

The minimum and maximum percentages of tension reinforcement need to be checked.

The design of the stirrups can be carried out using Table 62 of SP16.

The minimum shear reinforcement needs to be checked, as per clause 26.5.1.6 of IS 456, 2000:

$$A_{sv}/b\ S_v \geq 0.4/(0.87\ f_y)$$

where

A_{sv} = the total cross-sectional area of the stirrup legs effective in shear
S_v = the stirrup spacing along the length of the member
b = the breadth of the beam or the breadth of the web of a flanged beam
f_y = the characteristic strength of stirrup reinforcement in N/mm^2

Design of Short-Frame Beams (4–4)

Detailed design calculations for shorter frame beams have done in tabular form, and are shown in Tables 4.74a and 4.74b.

Design of Longer-Frame Beams (B–B)

Detailed design calculations for longer frame beams have done in tabular form, and are shown in Tables 4.74c and 4.74d.

TABLE 4.74A
Design of shorter frame beams (4–4) for longitudinal reinforcements (see Tables 4.70 and 4.71)

Beam marked	Factored bending moment (M_u) kNm	Factored shear force (V_u) kN	Grade of concrete N/m²	Grade of steel N/m²	Dimensions of the beam b (mm)	D (mm)	Effective depth(d) d(mm)	Effective depth/effective cover (d'/d)	M_u/bd^2	M_{ulim}/bd^2 Table C of SP16	As per table 3 / 55 of SP16 p_t	p_c	As per clause 26.5.1.1 of IS 456: minimum tension reinforcement Minimum percentage of steel, $p_t = 85/f_y$	Maximum: $p_t = 4\%$	Area of tension face A_{st} (mm²)	Area of compression face A_{sc} (mm²)
Beam AB																
Support A	−345.0	200.10	25	500	250	550	504	0.1	5.43	3.325	1.482	0.583	0.17	4	1,867.32	734.58
Span AB	97.005		25	500	250	550	504	0.1	1.53	3.325	0.387	0	0.17	4	487.62	0
Support B	−349.2	201.60	25	500	250	550	504	0.1	5.50	3.325	1.500	0.603	0.17	4	1,890.00	759.78
Beam BC																
Support B	−279.0	185.25	25	500	250	550	504	0.1	4.39	3.325	1.219	0.298	0.17	4	1,535.94	375.48
Span BC	26.7		25	500	250	550	504	0.1	0.42	3.325	0.106	0	0.17	4	214.20	0
Support C	−279.0	185.25	25	500	250	550	504	0.1	4.39	3.325	1.219	0.298	0.17	4	1,535.94	375.48
Beam CD																
Support C	−348.9	201.45	25	500	250	550	504	0.1	5.49	3.325	1.500	0.603	0.17	4	1,890.00	759.78
Span CD	97.005		25	500	250	550	504	0.1	1.53	3.325	0.387	0	0.17	4	487.62	0
Support D	−344.7	199.95	25	500	250	550	504	0.1	5.43	3.325	1.482	0.583	0.17	4	1,867.32	734.58

TABLE 4.74B
Design of shear reinforcements of shorter frame beams (4–4)

Beam marked	Design shear strength of concrete as per Table 61 of SP16 (N/mm²)	V_{us} (kN)	Spacing of stirrups required, $S_v = 0.87f_y.A_{sv}.d/V_{us}$, as per clause 4.3 of SP16, considering 8Φ2L stirrups (mm)	Check for minimum shear reinforcement, $S_v = 0.87f_y.A_{sv}/(0.4b)$, $S_v = 0.75d$ and $S_v = 300$, whichever is less, as per clauses 26.5.1.5 and 26.5.1.5 of IS 456 (mm)			Spacing of stirrups provided (mm)
				Minimum shear reinforcement criteria (mm)	0.75d (mm)	300 mm	
Beam AB							
Support A	0.736	$V_{us} = (200.1) - (0.736 \times 250 \times 504/1000) = 107.364$ kN	204	435	378	300	300
Span AB	0.370	–46.62		435	378	300	175
Support B	0.740	108.36	202	435	378	300	175
Beam BC							
Support B	0.690	98.31	223	435	378	300	200
Span BC	0.330	–41.58		435	378	300	300
Support C	0.690	98.31	223	435	378	300	200
Beam CD							
Support C	0.740	108.21	203	435	378	300	175
Span CD	0.370	–46.62		435	378	300	300
Support D	0.736	107.21	204	435	378	300	175

TABLE 4.74C

Design of longer frame beams (B–B) for longitudinal reinforcements (see Tables 4.72 and 4.73)

Beam marked	Factored bending moment (M_u)	Factored shear force (V_u)	Grade of concrete (f_ck)	Grade of steel (f_y)	Dimensions of the beam b (mm)	Dimensions of the beam D (mm)	Effective depth D (mm)	Effective depth/effective cover (d'/d)	M_u/bd^2	Table C of SP16 M_{ulim}/bd^2	Percentage of steel, as per table 3/55 of SP16 p_t (%)	p_c (%)	As per clause 26.5.1.1 of IS 456: minimum tension reinforcement — Minimum Percentage of steel $p_t = 85/f_y$ (%)	Maximum Percentage of steel $p_t = 4\%$	Area of tension face A_{st} (mm²)	Area of compression face A_{sc} (mm²)
Beam1–2	kNm	kN	N/mm²	N/mm²					N/mm²							
Support 1	−415.2	235.65	25	500	250	650	604	0.1	4.55	3.325	1.257	0.34	0.17	4	1,898.07	513.4
Span 1–2	240		25	500	250	650	604	0.1	2.63	3.325	0.711	0	0.17	4	1,073.61	0
Support 2	−487.05	268.8	25	500	250	650	604	0.1	5.34	3.325	1.459	0.559	0.17	4	2,203.09	844.09
Beam 2–3																
Support 2	−328.2	182.85	25	500	250	650	604	0.1	3.6	3.325	1.014	0.076	0.17	4	1,531.14	114.76
Span 2–3	123.3		25	500	250	650	604	0.1	1.35	3.325	0.333	0	0.17	4	502.83	0
Support 3	−283.5	166.8	25	500	250	650	604	0.1	3.11	3.325	0.866	0	0.17	4	1,307.66	0
Beam 3–4																
Support 3	−328.2	189.45	25	500	250	650	604	0.1	3.6	3.325	1.014	0.076	0.17	4	1,531.14	114.76
Span 3–4	119.1		25	500	250	650	604	0.1	1.31	3.325	0.33	0	0.17	4	498.3	0
Support 4	−343.5	195	25	500	250	650	604	0.1	3.77	3.325	1.057	0.124	0.17	4	1,596.07	187.24
Beam 4–5																
Support 4	−348.9	196.5	25	500	250	650	604	0.1	3.83	3.325	1.073	0.14	0.17	4	1,620.23	211.4
Span 4–5	133.95		25	500	250	650	604	0.1	1.47	3.325	0.366	0	0.17	4	552.66	0
Support 5	−355.8	198.6	25	500	250	650	604 −46	0.1	3.9	3.325	1.091	0.159	0.17	4	1,647.41	240.09
Beam 5–6																
Support 5	−321.75	187.05	25	500	250	650	604	0.1	3.53	3.325	0.997	0.056	0.17	4	1,505.47	84.56
Span 5–6	163.2		25	500	250	650	604	0.1	1.79	3.325	0.456	0	0.17	4	688.56	0
Support 6	−321.75	187.05	25	500	250	650	604	0.1	3.53	3.325	0.997	0.056	0.17	4	1,505.47	84.56
Beam 6–7																
Support 6	−420.9	219.15	25	500	250	650	604	0.1	4.61	3.325	1.273	0.355	0.17	4	1,922.23	536.05
Span 6–7	124.2		25	500	250	650	604	0.1	1.36	3.325	0.346	0	0.17	4	522.46	0
Support 7	−352.8	196.5	25		250	650	604	0.1	3.87	3.325	1.08	0.15	0.17	4	1,630.8	226.5

TABLE 4.74D
Design of shear reinforcements of longer frame beams (B–B)

Beam marked	Design shear strength of concrete as per Table 61 of SP16 (N/mm²)	V_{us} (kN)	Spacing of stirrups required, $S_v = 0.87f_y \cdot A_{sv} \cdot d/V_{us}$, as per clause 4.3 of SP16, considering 8Φ2L stirrups (mm)	Check for minimum shear reinforcement, $S_v = 0.87f_y \cdot A_{sv}/(0.4b)$, $S_v = 0.75d$ and $S_v = 300$, whichever is less, as per clauses 26.5.1.5 and 26.5.1.5 of IS 456 (mm)			Spacing of stirrups provided (mm)
				Minimum shear reinforcement criteria (mm)	0.75d (mm)	300 mm	
Beam 1–2							
Support 1	0.7	129.95	202	435	453	300	175
Span 1–2	0.49			435	453	300	300
Support 2	0.732	158.268	166	435	453	300	150
Beam 2–3							
Support 2	0.64	86.21	305	435	453	300	250
Span 2–3	0.33			435	453	300	300
Support 3	0.61	74.69	352	435	453	300	300
Beam 3–4							
Support 3	0.64	92.81	283	435	453	300	250
Span 3–4	0.33			435	453	300	300
Support 4	0.65	96.85	271	435	453	300	250
Beam 4–5							
Support 4	0.654	97.746	269	435	453	300	250
Span 4–5	0.348			435	453	300	300
Support 5	0.66	98.94	266	435	453	300	250
Beam 5–6							
Support 5	0.64	90.41	291	435	453	300	250
Span 5–6	0.378			435	453	300	300
Support 6	0.64	90.41	291	435	453	300	250
Beam 6–7							
Support 6	0.705	112.695	233	435	453	300	200
Span 6–7	0.36			435	453	300	300
Support 7	0.655	97.595	269	435	453	300	250

Detailing of Beam

For the detailing of the beam, the following points need to be maintained, as per SP 34, while showing the detail of the reinforcements:

1. the bars are to be symmetrically placed about the vertical center line of the beams
2. when there are only two bars in a row, these shall be placed at the outer edges
3. when bars of different diameters are placed in a single bottom row, the larger-diameter bars need to be placed on the outer side
4. when bars in different horizontal rows have different diameters, the larger-diameter bars are to be placed in the bottom row

The minimum horizontal distance between longitudinal bars is not to be less than the maximum of the following recommendations:

1. the diameter of the longitudinal bars, if the diameters are equal
2. the diameter of the larger bar, if the diameters are unequal
3. 5 mm more than the nominal maximum size of coarse aggregate

However, in general and in practice, a minimum 50 to 75 mm clear gap between longitudinal reinforcements needs to be maintained to avoid any honeycombing during casting and to have a properly compacted concrete.

Reinforcement detailing in the form of working drawings are shown in Figures 4.42 and 4.43.

4.4.13 Design of Columns

Let us consider the column marked B4 at the ground floor level (refer to Figure 4.38).

The positioning of the column to be designed is shown in Figure 4.38.

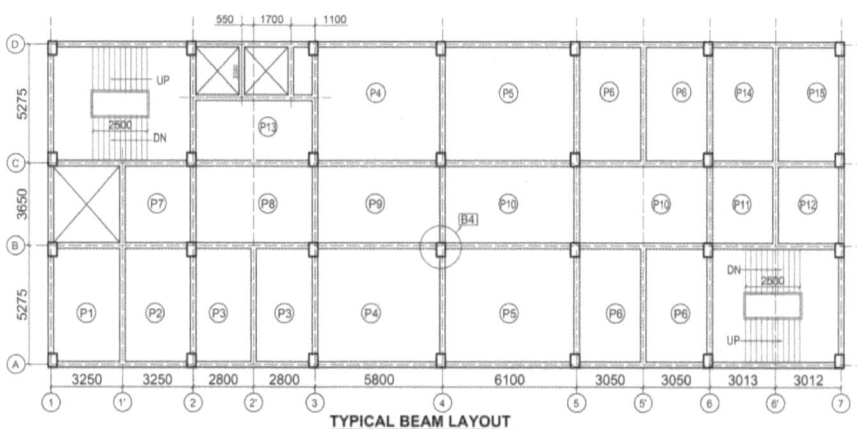

TYPICAL BEAM LAYOUT

FIGURE 4.38 Position of column B4

Reference to the design values of the different parameters is made as follows: Tables 4.27/4.55, 4.28/4.56, 4.29/4.53, 4.30/4.54, 4.36/4.59, 4.38/4.57, 4.39/4.58, 4.41, 4.42/4.61, 4.43/4.62, 4.44, 4.45/4.63, 4.46/4.64, 4.47, 4.48/4.65, 4.49/4.66, 4.50, 4.51/4.67, 4.52/4.68.

The values of P_u, M_{ux} and M_{uy}, for different load cases, have been computed and are shown in Table 4.75.

Loads Due to Dead Loads and Live Loads

With respect to the shear force calculation of the shorter frame (4–4) at ground floor level (refer to Tables 4.30/4.54, following the encircled values in Table 4.30, and Figure 4.19c).

In span AB, the shear force at the left of B4 = ⟨109.2⟩kN

In span BC, the shear force at the right of B4 = ⟨54.8⟩kN

Therefore, the axial force on the column marked B4 at ground floor level

$$= 109.2 + 54.8 = ⟨164⟩ \text{ kN (compressive)}$$

The bending moment due to dead and live loads on the column at ground floor level = ⟨-18⟩ kNm

(refer to Figure 4.19b and follow the encircled values in Table 4.54)

Similarly, refer to Tables 4.39/4.58 and Figure 4.27c for the longer frame analysis (B–B) at ground floor level.

In span 3–4, the shear force at the left of B4 = ⟨118.1⟩ kN

In span 4–5, the shear force at the right of B4 = ⟨124⟩kN

(refer to Figure 4.27c and follow the encircled values in Table 4.39)

The axial force on the column marked B4 at ground floor level

$$= 118.1 + 124 = ⟨242.1⟩ \text{ kN (compressive)}$$

The bending moment due to the dead and live loads on the column at ground floor level = ⟨-3.85⟩kNm

(refer to Figure 4.27b and follow the encircled values)

Loads Due to Dead Loads Only

With respect to the shear force calculation of the longer frame analysis (B–B) at ground floor level (refer to Tables 4.28/4.56, following the encircled values in Table 4.28, and Figure 4.22c).

In span AB, the shear force at the left of B4 = ⟨74.2⟩ kN

In span BC, the shear force at the right of B4 = ⟨36.7⟩ kN

The axial force on the column marked B4 at ground floor level

$$= 74.2 + 36.7 = ⟨110.9⟩ \text{ kN (compressive)}$$

The bending moment due to dead loads on the column at ground floor level = -7.51 kNm

(refer to Figure 4.22b and follow the encircled values)

Similarly, refer to Tables 4.37/4.60, following the encircled values in Table 4.37, and Figure 4.32c for the longer frame analysis.

In span 3–4, the shear force at the left of B4 = 84 kN

In span 4–5, the shear force at the right of B4 = 87.7 kN

The axial force on the column marked B4 at ground floor level = 84 + 87.7 = 171.7 kN

The bending moment due to dead loads on the column at ground floor level = -1.1 kNm

(refer to Figure 4.32b and follow the encircled values)

Loads Due to Wind Loads

With respect to the wind load analysis of the shorter frame (4–4) and values considered at ground floor level (refer to Tables 4.43/4.62 and Table 4.40).

The axial force on the column marked B4 = ±20.2 kN

The bending moment in column B4 = ±69.3 kNm

With respect to the wind load analysis of the longer frame (B–B) at ground floor level (refer to Tables 4.46/4.64, following the encircled values).

The axial force on the column marked B4 = ±0.97 kN

The bending moment in column B4 = ±15.98 kNm

Loads Due to Seismic Loads

With respect to the seismic load analysis of the shorter frame (4–4) and values considered at ground floor level (refer to Tables 4.49/4.66, following the encircled values in the tables).

The axial force on the column marked B4 = ±141.3 kN

The bending moment in column B4 = ±270.9 kNm

With respect to the seismic load analysis of the longer frame (refer to Tables 4.52/4.68, following the encircled values).

The axial force on the column marked B4 = ±12.1 kN

The bending moment in column B4 = ±268 kNm

TABLE 4.75
Load combination for design of columns at ground floor level

Load cases	Combination of loads with partial safety factor of loads	P_u (kN)	M_{ux} (kNm)	M_{uy} (kNm)
1	1.5(DL + LL)	*3,470.2875	−5.775	−34.5
2	1.5(DL + WL)	**2,529.368	25.62	92.685
3	1.5(DL − WL)	2,465.858	−22.32	−115.215
4	1.5(DL + EQ)	2,303.8125	403.65	395.085
5	1.5(DL − EQ)	2,691.4125	−400.35	−417.615
6	1.2(DL + LL + WL)	2,753.154	14.34	55.56
7	1.2(DL + LL − WL)	2,799.306	−23.58	−110.76
8	1.2(DL + LL + EQ)	2,621.19	316.98	297.48
9	1.2(DL + LL − EQ)	2,931.27	−326.22	−352.68
10	0.9DL + 1.5EQ	1,315.845	402.99	399.591
11	0.9DL − 1.5EQ	1,703.445	−401.01	−413.109

* Sample Calculation for Load Case 1

The building has five typical floors and a roof. The vertical gravity forces due to the dead and live loads for the roof need to be calculated separately, but, for this exercise, 50 percent of a typical floor load is considered to make it more straightforward, as there is no full brickwork load on the roof, etc. However, the self-weight of the column has been included separately.

The self-weight of the columns = $21.1 \times 0.3 \times 0.7 \times 25 = 110.775$ kN

$$P_u = 1.5 \times ((\;\boxed{164} + \boxed{242.1}\;) \times 5.5) + 110.775 = 3,470.28 \text{ kN}$$

*Encircled values are already calculated earlier

$$M_{ux} = 1.5 \times -3.85 = -5.775 \text{ kNm}$$
$$M_{uy} = 1.5 \times \boxed{-18} = -27 \text{ kNm}$$

(refer to Figure 4.19b and follow the encircled values)

Sample Calculation for Load Case 2

$$P_u = ((110.9 + 171.7) \times 5.5 + (21.1 \times 0.3 \times 0.7 \times 25) + (20.2 + 0.97)) \times 1.5 = 2,529.36 \text{ kN}$$

$$M_{ux} = 1.5 \times (1.1 + 15.98) = 25.62 \text{ kNm}$$

$$M_{uy} = 1.5 \times (-7.51 + 69.3) = 92.685 \text{ kNm}$$

TABLE 4.76A
Calculation of final design values of P_u, M_{ux} and M_{uy} for column B_4 (first trial)

Step -1

Load Case	Column dimension					Eccentricity Check as per Clause no. 25.4 of IS 456				Final Factored moment	
	b_x	d_y	P_u	M_{ux}	M_{uy}	e_x* min	e_y** min	$e_x = M_{ux}/P_u$	$e_y = M_{uy}/P_u$	M_{ux}	M_{uy}
	mm	mm	kN	kNm	kNm	mm	mm	mm	mm	kNm	kNm
1	300	700	3,470.2875	-5.775	-27	20	28.63	1.66	7.78	-69.41	99.35
2	300	700	2,527.9125	25.62	-11.265	20	28.63	10.13	4.46	-50.56	72.37
3	300	700	2,467.3125	-22.32	-11.265	20	28.63	9.05	4.57	-49.35	70.64
4	300	700	2,499.0675	1.65	92.685	20	28.63	0.66	37.09	-49.98	92.685
5	300	700	2,496.1575	1.65	-115.21	20	28.63	0.66	46.16	-49.92	-115.215
6	300	700	2,285.6625	403.65	-11.265	20	28.63	176.6	4.93	-403.65	65.44
7	300	700	2,691.4125	-400.35	-11.265	20	28.63	148.75	4.19	400.35	77.06
8	300	700	2,515.7625	1.65	395.085	20	28.63	0.66	157.04	-50.32	395.085
9	300	700	2,479.4625	1.65	-417.61	20	28.63	0.67	168.43	-49.59	-417.615
10	300	700	2,751.99	14.34	-27.6	20	28.63	5.21	10.03	-55.04	78.79
11	300	700	2,800.47	-23.58	-27.6	20	28.63	8.42	9.86	-56.01	80.18
12	300	700	2,777.394	-4.62	55.56	20	28.63	1.66	20	-55.55	79.52
13	300	700	2,775.066	-4.62	-110.76	20	28.63	1.66	39.91	-55.5	-110.76
14	300	700	2,606.67	316.98	-27.6	20	28.63	121.6	10.59	-316.98	74.63
15	300	700	2,945.79	-326.22	-27.6	20	28.63	110.74	9.37	326.22	84.34
16	300	700	2,790.75	-4.62	297.48	20	28.63	1.66	106.6	-55.82	297.48
17	300	700	2,761.71	-4.62	-352.68	20	28.63	1.67	127.7	-55.23	-352.68
18	300	700	1,297.695	402.99	-6.759	20	28.63	310.54	5.21	-402.99	37.15
19	300	700	1,721.595	402.99	-6.759	20	28.63	234.08	3.93	-402.99	49.29
20	300	700	1,527.795	0.99	399.591	20	28.63	0.65	261.55	-30.56	399.591
21	300	700	1,491.495	0.99	-413.11	20	28.63	0.66	276.98	-29.83	-413.109

Notes: *e_x min = 2,650/500 + 300/30 = 15.3 mm < 20 mm, so e_x to be considered = 20 mm. **e_y min = 2,650/500 + 700/30 = 28.63 mm > 20 mm, so e_y to be considered = 28.63 mm. *** $M_{ux} = P_u \times e_x$. **** $M_{uy} = P_u \times e_y$.

TABLE 4.76B
Design calculations for column B_4 (first trial)

Step 2a

Load case	P_u	M_{ux}	M_{uy}	Assumed value Percentage of steel p	p/f_{ck}	d'	d'/D	$P_u/f_{ck}bD$	Chart 50 of SP16 $M_{uxl}/f_{ck}bD^2$ k fromP–M curve	$M_{uxl}=kf_{ck}bD^2$	Is $M_{uxl} > M_{ux}$?
	kN	kNm	kNm	p		Assumed					
1	3,470.2875	−69.41	99.35	2	0.08		0.2	0.66	0.03	47.25	Fail
2	3,470.2875	−69.41	99.35	2	0.08		0.2	0.66	0.03	47.25	Fail
3	2,527.9125	−50.56	72.37	2	0.08		0.2	0.48	0.08	126	Pass
4	2,467.3125	−49.35	70.64	2	0.08	Nominal cover =40mm Link dia. 8mm Longitudinal bar dia. 20mm $d' = 40+8+10$ = 58 mm	0.2	0.47	0.08	126	Pass
5	2,499.0675	−49.98	92.685	2	0.08		0.2	0.48	0.08	126	Pass
6	2,496.1575	−49.92	— (115.215)	2	0.08		0.2	0.48	0.08	126	Pass
7	2,285.6625	−403.65	65.44	2	0.08		0.2	0.44	0.085	133.875	Fail
8	2,691.4125	400.35	77.06	2	0.08		0.2	0.51	0.075	118.125	Fail
9	2,515.7625	−50.32	395.085	2	0.08		0.2	0.48	0.08	126	Pass
10	2,479.4625	−49.59	— (417.615)	2	0.08		0.2	0.47	0.08	126	Pass
11	2,751.99	−55.04	78.79	2	0.08		0.2	0.52	0.075	118.125	Pass
12	2,800.47	−56.01	80.18	2	0.08		0.2	0.53	0.07	110.25	Pass
13	2,777.394	−55.55	79.52	2	0.08		0.2	0.53	0.07	110.25	Pass

14	2,775.066	−55.5	−110.76	2	0.08	0.2	0.53	0.07	110.25	Pass
15	2,606.67	−316.98	74.63	2	0.08	0.2	0.5	0.075	118.125	Fail
16	2,945.79	326.22	84.34	2	0.08	0.2	0.56	0.06	94.5	Fail
17	2,790.75	−55.82	297.48	2	0.08	0.2	0.53	0.07	110.25	Pass
18	2,761.71	−55.23	−352.68	2	0.08	0.2	0.53	0.07	110.25	Pass
19	1,297.695	−402.99	37.15	2	0.08	0.2	0.25	0.11	173.25	Fail
20	1,721.595	−402.99	49.29	2	0.08	0.2	0.33	0.105	165.375	Fail
21	1,527.795	−30.56	399.591	2	0.08	0.2	0.29	0.108	170.1	Pass

TABLE 4.76C
Design calculations for column B4 (first trial)

Step 2b

Load case	P_u kN	M_{ux} kNm	M_{uy} kNm	Assume Percentage of steel p	Uniaxial moment capacity about y–y p/f_{ck}	d'	d'/D	$Pu/f_{ck}bD$	Chart 50 of SP16 $M_{uy1}/f_{ck}bD^2$ (k)	$M_{uy1} = k f_{ck}bD^2$	Is $M_{uy1} > M_{uy}$?
1	3,470.2875	−69.41	99.35	2	0.08		0.1	0.66	0.03	110.25	Pass
2	3,470.2875	−69.41	99.35	2	0.08		0.1	0.66	0.03	110.25	Pass
3	2,527.9125	−50.56	72.37	2	0.08	Nominal cover	0.1	0.48	0.08	294	Pass
4	2,467.3125	−49.35	70.64	2	0.08	=40mmlink	0.1	0.47	0.08	294	Pass
5	2,499.0675	−49.98	92.685	2	0.08	diameter,8mm	0.1	0.48	0.08	294	Pass
6	2,496.1575	−49.92	−115.215	2	0.08	main bar,	0.1	0.48	0.08	294	Pass
7	2,285.6625	−403.65	65.44	2	0.08	20mm	0.1	0.44	0.085	312.375	Pass
8	2,691.4125	400.35	77.06	2	0.08	(assumed)	0.1	0.51	0.075	275.625	Pass
9	2,515.7625	−50.32	395.085	2	0.08	d' = 40+8+10	0.1	0.48	0.08	294	Fail
10	2,479.4625	−49.59	−417.615	2	0.08	=58mm	0.1	0.47	0.08	294	Fail
11	2,751.99	−55.04	78.79	2	0.08		0.1	0.52	0.075	275.625	Pass
12	2,800.47	−56.01	80.18	2	0.08		0.1	0.53	0.07	257.25	Pass
13	2,777.394	−55.55	79.52	2	0.08		0.1	0.53	0.07	257.25	Pass
14	2,775.066	−55.5	−110.76	2	0.08		0.1	0.53	0.07	257.25	Pass
15	2,606.67	−316.98	74.63	2	0.08		0.1	0.5	0.075	275.625	Pass
16	2,945.79	326.22	84.34	2	0.08		0.1	0.56	0.06	220.5	Pass
17	2,790.75	−55.82	297.48	2	0.08		0.1	0.53	0.07	257.25	Fail
18	2,761.71	−55.23	−352.68	2	0.08		0.1	0.53	0.07	257.25	Fail
19	1,297.695	−402.99	37.15	2	0.08		0.1	0.25	0.11	404.25	Pass
20	1,721.595	−402.99	49.29	2	0.08		0.1	0.33	0.105	385.875	Pass
21	1,527.795	−30.56	399.591	2	0.08		0.1	0.29	0.108	396.9	Fail

Similarly, for other load cases, the values of P_u, M_{ux} and M_{uy} have been computed, and they are shown in Table 4.75.

Before proceeding to the final design of the columns, first the eccentricity of the column has to be checked. As per clause 25.4 of IS 456, 2000, all the columns need to be designed with minimal eccentricity. Therefore, if the eccentricity of the column is greater than that of said minimum value, the bending moment calculated from analysis needs to be adopted; otherwise, the column has to be designed with a bending moment generated by multiplying the vertical load by the minimum eccentricity. Minimum eccentricity has been calculated as per the IS 456, 2000, recommendation, given below:

Minimum eccentricity = unsupported length of column / 500 + lateral
dimension / 30

Subject to a minimum of 20 mm

In analysis, the column size assumed is 300 mm × 700 mm

The unsupported length of the column is as per clause 25.1.3b of IS 456, 2000: it is to be the clear distance between the floor and the underside of the shallower beam framing into the columns in each direction at the next higher floor level, for a framed building.

For this building, the unsupported length of the column = 3,200 − 550 = 2,650 mm

With reference to Table 4.75, showing P_u, M_{ux} and M_{uy} for different load combinations, the first trial of designing column B4 is shown in Tables 4.76a, 4.76b and 4.76c.

*e_x min = 2650/500 + 300/30 = 15.3 mm < 20 mm, so, e_x min = 20 mm

**e_y min = 2650/500 + 700/30 = 28.63 mm > 20 mm, so, e_y min = 28.63 mm

The column which to be designed is subjected to bending in both the axis, so it is to be designed as member with axial load and biaxial bending. As per Clause no. 39.6 of IS 456 for members subjected to combined axial load and biaxial bending should satisfy following condition:

$$[M_{ux}/M_{ux1}]^{\alpha n} + [M_{uy}/M_{uy1}]^{\alpha n} \leq 1.0$$

Where,

M_{ux}, M_{uy} = Bending moments about x and y axes due to design loads,
M_{ux1}, M_{uy1} = maximum uniaxial bending moment capacity for an axial load of Pu bending about x and y axes respectively.
α_n is an exponent whose value depends on P_u/P_{uz}

P_u/P_{uz}	α_n
≤ 0.2	1.0
≥ 0.8	2.0

Now, to calculate the uniaxial moment capacity of section for axial load P_u following steps may be followed:

TABLE 4.77A
Calculation of final design values of P_u, M_{ux} and M_{uy} for column B_4

Step -1

Load Case	Column dimension					Eccentricity check as per Clause no. 25.4 of IS 456				Final Factored moment	
	b_x	d_y	P_u	M_{ux}	M_{uy}	e_x* min	e_y* min	$e_x = M_{ux}/P_u$	$e_y = M_{uy}/P_u$	***M_{ux}	****M_{uy}
	mm	mm	kN	kNm	kNm	mm	mm	mm	mm	kNm	kNm
1	500	700	3,581.06	−5.775	−27	20	28.63	1.61	9.63	−71.62	102.53
2	500	700	2,638.688	25.62	−11.265	20	28.63	9.71	4.27	−52.77	75.55
3	500	700	2,578.088	−22.32	−11.265	20	28.63	8.66	4.37	−51.56	73.81
4	500	700	2,609.843	1.65	92.685	20	28.63	0.63	35.51	−52.2	92.685
5	500	700	2,606.933	1.65	−115.215	20	28.63	0.63	44.2	−52.14	−115.215
6	500	700	2,396.43	403.65	−11.265	20	28.63	168.44	4.7	−403.65	68.61
7	500	700	2,802.18	−400.35	−11.265	20	28.63	142.87	4.02	400.35	80.23
8	500	700	2,626.53	1.65	395.085	20	28.63	0.63	150.42	−52.53	395.085
9	500	700	2,590.23	1.65	−417.615	20	28.63	0.64	161.23	−51.8	−417.615
10	500	700	2,840.61	14.34	−27.6	20	28.63	5.05	9.72	−56.81	81.33
11	500	700	2,889.09	−23.58	−27.6	20	28.63	8.16	9.55	−57.78	82.71
12	500	700	2,866.014	−4.62	55.56	20	28.63	1.61	19.39	−57.32	82.05
13	500	700	2,863.686	−4.62	−110.76	20	28.63	1.61	38.68	−57.27	−110.76
14	500	700	2,695.29	316.98	−27.6	20	28.63	117.61	10.24	−316.98	77.17
15	500	700	3,034.41	−326.22	−27.6	20	28.63	107.51	9.1	326.22	86.88
16	500	700	2,879.37	−4.62	297.48	20	28.63	1.6	103.31	−57.59	297.48
17	500	700	2,850.33	−4.62	−352.68	20	28.63	1.62	123.73	−57.01	−352.68
18	500	700	1,371.545	402.99	−6.759	20	28.63	293.82	4.93	−402.99	39.27
19	500	700	1,795.445	402.99	−6.759	20	28.63	224.45	3.76	−402.99	51.4
20	500	700	1,601.645	0.99	399.591	20	28.63	0.62	249.49	−32.03	399.591
21	500	700	1,565.345	0.99	−413.109	20	28.63	0.63	263.91	−31.31	−413.109

Notes: *e_x min = 2650/500 + 500/30 = 21.97 mm = 21.97 mm > 20 mm, so e_x min = 21.97 mm. **e_y min = 2650/500 + 700/30 = 28.63 mm = 28.63 mm > 20 mm, so e_y min = 28.63 mm. ***M_{ux} = P_u x e_x. ****M_{uy} = P_u x e_y.

TABLE 4.77B
Final design calculations for column B4

Step 2a

Load case	P_u kN	M_{ux} kNm	M_{uy} kNm	Assumed Percentage of steel p	p/f_{ck}	d' Assumed	d'/D	$P_u/f_{ck}bD$	k fromP–M curve	$M_{ux1} = k f_{ck}bD^2$	Is Mux1 > Mux?
1	3,581.0625	−71.62	102.53	1.2	0.05	Nominal	0.1	0.41	0.08	350	Pass
2	2,638.6875	−52.77	75.55	1.2	0.05	cover =	0.1	0.3	0.1	437.5	Pass
3	2,578.0875	−51.56	73.81	1.2	0.05	40 mm link	0.1	0.29	0.1	437.5	Pass
4	2,609.8425	−52.2	92.685	1.2	0.05	diameter,	0.1	0.3	0.1	437.5	Pass
5	2,606.9325	−52.14	−115.215	1.2	0.05	8 mm	0.1	0.3	0.1	437.5	Pass
6	2,396.4375	−403.65	68.61	1.2	0.05	longitudinal	0.1	0.27	0.11	481.25	Pass
7	2,802.1875	400.35	80.23	1.2	0.05	bar	0.1	0.32	0.095	415.625	Pass
8	2,626.5375	−52.53	395.085	1.2	0.05	diameter,	0.1	0.3	0.1	437.5	Pass
9	2,590.2375	−51.8	−417.615	1.2	0.05	20 mm	0.1	0.3	0.1	437.5	Pass
10	2,840.61	−56.81	81.33	1.2	0.05	d' = 40 + 8 +	0.1	0.32	0.095	415.625	Pass
11	2,889.09	−57.78	82.71	1.2	0.05	10 = 58 mm	0.1	0.33	0.09	393.75	Pass
12	2,866.014	−57.32	82.05	1.2	0.05		0.1	0.33	0.09	393.75	Pass
13	2,863.686	−57.27	−110.76	1.2	0.05		0.1	0.33	0.09	393.75	Pass
14	2,695.29	−316.98	77.17	1.2	0.05		0.1	0.31	0.1	437.5	Pass
15	3,034.41	326.22	86.88	1.2	0.05		0.1	0.35	0.085	371.875	Pass
16	2,879.37	−57.59	297.48	1.2	0.05		0.1	0.33	0.09	393.75	Pass
17	2,850.33	−57.01	−352.68	1.2	0.05		0.1	0.33	0.09	393.75	Pass
18	1,371.545	−402.99	39.27	1.2	0.05		0.1	0.16	0.11	481.25	Pass
19	1,795.445	−402.99	51.4	1.2	0.05		0.1	0.21	0.105	459.375	Pass
20	1,601.645	−32.03	399.591	1.2	0.05		0.1	0.18	0.11	481.25	Pass
21	1,565.345	−31.31	−413.109	1.2	0.05		0.1	0.18	0.11	481.25	Pass

The columns under "Uniaxial moment capacity about x–x" are p/fck, d', d'/D, Pu/fckbD. The columns under "Chart 48 of SP16" are k fromP–M curve, Mux1 = k fck bD².

TABLE 4.77C
Final design calculations for column B4

Step 2b

Load case	P_u kN	M_{ux} kNm	M_{uy} kNm	Assumed Percentage of steel p	Uniaxial moment capacity about y–y p/f_{ck}	d'	Chart 48 of SP16 d'/D	$P_u/f_{ck}bD$	Muy1/fckbD2 k	Muy1 = k.fckbD2	IsMuy1 > Muy?
1	3,581.0625	−71.62	102.53	1.2	0.05	Nominal	0.1	0.41	0.08	490	Pass
2	2,638.6875	−52.77	75.55	1.2	0.05	cover	0.1	0.3	0.1	612.5	Pass
3	2,578.0875	−51.56	73.81	1.2	0.05	=40mmlink	0.1	0.29	0.1	612.5	Pass
4	2,609.8425	−52.2	92.685	1.2	0.05	diameter,	0.1	0.3	0.1	612.5	Pass
5	2,606.9325	−52.14	−115.215	1.2	0.05	8mm main	0.1	0.3	0.1	612.5	Pass
6	2,396.4375	−403.65	68.61	1.2	0.05	bar, 20mm	0.1	0.27	0.11	673.75	Pass
7	2,802.1875	400.35	80.23	1.2	0.05	(assumed)	0.1	0.32	0.095	581.875	Pass
8	2,626.5375	−52.53	395.085	1.2	0.05	d' = 40+8+10	0.1	0.3	0.1	612.5	Pass
9	2,590.2375	−51.8	−417.615	1.2	0.05	=58mm	0.1	0.3	0.1	612.5	Pass
10	2,840.61	−56.81	81.33	1.2	0.05		0.1	0.32	0.095	581.875	Pass
11	2,889.09	−57.78	82.71	1.2	0.05		0.1	0.33	0.09	551.25	Pass
12	2,866.014	−57.32	82.05	1.2	0.05		0.1	0.33	0.09	551.25	Pass
13	2,863.686	−57.27	−110.76	1.2	0.05		0.1	0.33	0.09	551.25	Pass
14	2,695.29	−316.98	77.17	1.2	0.05		0.1	0.31	0.1	612.5	Pass
15	3,034.41	326.22	86.88	1.2	0.05		0.1	0.35	0.085	520.625	Pass
16	2,879.37	−57.59	297.48	1.2	0.05		0.1	0.33	0.09	551.25	Pass
17	2,850.33	−57.01	−352.68	1.2	0.05		0.1	0.33	0.09	551.25	Pass
18	1,371.545	−402.99	39.27	1.2	0.05		0.1	0.16	0.11	673.75	Pass
19	1,795.445	−402.99	51.4	1.2	0.05		0.1	0.21	0.105	643.125	Pass
20	1,601.645	−32.03	399.591	1.2	0.05		0.1	0.18	0.11	673.75	Pass
21	1,565.345	−31.31	−413.109	1.2	0.05		0.1	0.18	0.11	673.75	Pass

Calculate d'/D, where d' and D as described in IS 456, 2000

Corresponding to d'/D value and $f_y = 500$ N/mm², using charts 47 to 50 of SP16, M_{ux1} and M_{uy1} to be calculated.

From above calculation, it is observed that for different load cases, $M_{ux1} < M_{ux}$ & $M_{uy1} < M_{uy}$ with a high percentage of steel, so, the width of column need to be increased.

Let the width of column be 500 mm. In analysis the column width considered as 300 mm, so there will be some changes in forces and moments. But this changes are not made in designing columns and beams.

Final Design of Column B4

The final design calculations for the column are shown in Tables 4.77a, 4.77b, 4.77c and 4.77d.

Chart 63 of SP16 is used to calculate P_{uz}, whereas α_n is calculated as per clause 39.6 of IS 456, 2000, against the P_u/P_{uz} ratio. Biaxial checks have been made, and the calculations are furnished in Table 4.77d.

$$P_u/P_{uz} = 0.65 \text{ and, correspondingly, } \alpha_n = 1 + [(2 - 1)/(0.8 - 0.2)(0.65 - 0.2)] = 1.75$$

$$**(M_{ux}/M_{ux1})^{\alpha n} + (M_{uy}/M_{uy1})^{\alpha n} = (71.62/350)^{1.75} + (102.53/490)^{1.75} = 0.13$$

It has been observed that the maximum value of $[M_{ux}/M_{ux1}]^{\alpha n} + [M_{uy}/M_{uy1}]^{\alpha n} = 0.99 <$ Therefore, this column section may be provided.

The area of steel $= (1.2/100) \times (500 \times 700) = 4{,}200$ mm²

Let us therefore provide 14 no. 20T or bars.

Note that, as per clause 26.5.3.1 of IS 456, 2000, longitudinal reinforcements should satisfy the following requirements.

Minimum percentage of reinforcement $= 0.8\%$
Maximum percentage of reinforcement $= 6\%$
Minimum diameter of longitudinal bars $= 12$ mm

Design of Transverse Reinforcement/Lateral Ties

As per clause 26.5.3.2c of IS 456, 2000, the diameter of the polygonal links or lateral ties is to be not less than one-fourth of the diameter of the largest longitudinal bar (and in no case less than 16 mm).

The diameter of largest longitudinal bar is 20 mm. Therefore, the diameter of the lateral ties/links should be not be less than $20/4 = 5$ mm.

Let us provide 8 mm-diameter links, taking all other practical aspects into account.

The spacing/pitch of the lateral ties should be as per clause 26.5.3.2c of IS 456. The spacing/pitch should not to be less than the following.

TABLE 4.78
Final design calculations of combined check for column B4

Step 3

Load case	Factored design axial compression P_u	Percentage of steel	Gross area A_g	Chart 63 of SP16 P_{uz}/A_g	P_{uz}	P_u/P_{uz}	*α_n	As per clause 39.6 of IS 456	Combined check **$(M_{ux}/M_{ux1})^{\alpha n} + (M_{uy}/M_{uy1})^{\alpha n}$
	kN	%	mm²	N/mm²	kN				
1	3,581.0625	1.2	350,000	15.8	5,530	0.65	1.75		0.13
2	2,638.6875	1.2	350,000	15.8	5,530	0.48	1.47		0.09
3	2,578.0875	1.2	350,000	15.8	5,530	0.47	1.45		0.09
4	2,609.8425	1.2	350,000	15.8	5,530	0.47	1.45		0.11
5	2,606.9325	1.2	350,000	15.8	5,530	0.47	1.45		0.13
6	2,396.4375	1.2	350,000	15.8	5,530	0.43	1.38		0.83
7	2,802.1875	1.2	350,000	15.8	5,530	0.51	1.52		0.99
8	2,626.5375	1.2	350,000	15.8	5,530	0.47	1.45		0.58
9	2,590.2375	1.2	350,000	15.8	5,530	0.47	1.45		0.62
10	2,840.61	1.2	350,000	15.8	5,530	0.51	1.52		0.1
11	2,889.09	1.2	350,000	15.8	5,530	0.52	1.53		0.11
12	2,866.014	1.2	350,000	15.8	5,530	0.52	1.53		0.11
13	2,863.686	1.2	350,000	15.8	5,530	0.52	1.53		0.14
14	2,695.29	1.2	350,000	15.8	5,530	0.49	1.48		0.67
15	3,034.41	1.2	350,000	15.8	5,530	0.55	1.58		0.87
16	2,879.37	1.2	350,000	15.8	5,530	0.52	1.53		0.44
17	2,850.33	1.2	350,000	15.8	5,530	0.52	1.53		0.56
18	1,371.545	1.2	350,000	15.8	5,530	0.25	1.08		0.87
19	1,795.445	1.2	350,000	15.8	5,530	0.32	1.2		0.9
20	1,601.645	1.2	350,000	15.8	5,530	0.29	1.15		0.59
21	1,565.345	1.2	350,000	15.8	5,530	0.28	1.13		0.62

Note: *Calculation of α_n for load case 1.

The least lateral dimension of the compression members = 500 mm
Sixteen times the smallest diameter of the longitudinal reinforcement
bars = 16 × 20 = 320mm or 300mm, whichever is less.

As per clause 26.5.3.2c of IS 456, links may be 8T @ 300 mmc/c.
However, as per IS 13920, near beam–column junctions the spacing of the lateral ties should not be more than 75 mm, to satisfy the ductility requirement of a moment-resisting frame. This requirement of spacing is to be maintained for the portion L/4 from the face of the column in the beam and H/6 from the faces of the beams in the coloumns, where L = the clear span of the beam (i.e., the face to face distance

TABLE 4.79
Unfactored axial force (P) and bending moment (M_x and M_y) or different load combinations

Load case	Combination	P	Mx	My
1	(DL + LL)	2,313.525	−3.85	−23
2	(DL + WLX)	1,685.28	17.08	−7.51
3	(DL − WLX)	1,644.88	−14.88	−7.51
4	(DL + WLY)	1,666.05	1.1	61.79
5	(DL − WLY)	1,664.11	1.1	−76.81
6	(DL + EQX)	1,523.775	269.1	−7.51
7	(DL − EQX)	1,794.275	−266.9	−7.51
8	(DL + EQY)	1,677.175	1.1	263.39
9	(DL − EQY)	1,652.975	1.1	−278.41
10	(DL + LL + WLX)	2,293.325	11.95	−23
11	(DL + LL − WLX)	2,333.725	−19.65	−23
12	(DL + LL + WLY)	2,314.495	−3.85	46.3
13	(DL + LL − WLY)	2,312.555	−3.85	−92.3
14	(DL + LL + EQX)	2,172.225	264.15	−23
15	(DL + LL − EQX)	2,454.825	−271.85	−23
16	(DL + LL + EQY)	2,325.625	−3.85	247.9
17	(DL + LL + EQY)	2,301.425	−3.85	−293.9
18	DL + EQX	1,523.775	269.1	−7.51
19	DL − EQX	1,806.375	269.1	−7.51
20	DL + EQY	1,677.175	1.1	263.39
21	DL−EQY	1,652.975	1.1	−278.41

between columns) and H = the clear height of the column (i.e., the top face of the beam to the bottom face of the beam of the next floor level).

Arrangements of lateral ties to tie up all the longitudinal reinforcements are done as per the recommendations of IS 456.

Reinforcement detailing is shown in the form of a working drawing in Figure 4.44.

4.4.14 DESIGN OF FOUNDATIONS

Please note that in this problem we have designed one column and it's foundation as an example, and we have considered it as isolated footing. But when the total foundation system of all columns will be designed, then it may so happened that we have to consider combined footing with multiple columns or even raft foundation depending upon the load on the columns and safe bearing capacity of soil as per Geo-technical report.

To get the worst load case for design of footing, we may calculate the soil pressure, applying formula

$p = P/A \pm M_x / Z_x \pm M_y /Z_y$

P = Upward soil reaction
A = Area of footing
Mx, Mx = Bending moment
Zx, Zy = Section modulus

Design foundation loads are as given in Table 4.79:
Eccentricity Check – To check no tension condition of foundation $e \leq B/6$, Where,
B = Dimension of foundation in that direction
And $e = M/P$
M = Moment acting on footing, P = Axial load (Refer Table 4.73)
Soil pressure on foundation = $P/A \pm M/Z$ where P = Axial load,
A = Area of footing,
Z = Section modulus of footing
Critical section for one-way shear is considered as per Clause no. 32.2.4.1a of IS 456.
Two way shear check, has been made as per clause no. 31.6.1 of IS 456.
Moment of foundation slab to be calculated at the face of the column /pedestal, reinforcement requirement for the moment to be compared with the reinforcement requirement against shear, the higher one will be provided.

Design of Isolated rectangular footing
Design loads for foundation of column B4 as per Table 4.79 (Load case 15)
Axial compressive load including self weight of footing (10% of axial load on column)
$P_{total} = 2454.825 \times 1.1 = 2700.3$ kN
Bending Moment in x direction = $M_x = -271.85$ kNm
Bending Moment in z direction = $M_z = -23$ kNm

Geotechnical data:
Depth of foundation = 1.5 m
  Net safe bearing capacity = 190 kN/m² (from geotechnical report)
  Unit weight of soil = γ = 18 kN/m³ and N-value =15 (from geotechnical report)
  The design load case is dead load and seismic load combination, so as per IS 1893, 2016 clause 6.3.5.2 and Table 1, safe bearing capacity may be enhanced by 25%.
  So, permissible net safe bearing capacity = $190 \times 1.25 = 237.5$ kN/m²

General data:
Column size = 500 mm × 700 mm
Pedestal size = 700 mm × 900 mm

Area of footing required = Axial compressive force / Net safe bearing capacity of soil = 2700.3/ 237.5 = 11.4 m²
Considering, a footing size of 3.5 m × 4 m with area 14 m²

Check for eccentricity:

Eccentricity in x direction = M_x / P_{total} = 271.85 / 2700.3 = 0.1 m = 100 mm

Eccentricity in z direction = M_z / P_{total} = 23 / 2700.3 = 0.0085 m = 8.5 mm

Width of footing in x direction = 3500,

Allowable eccentricity = 3500/6 = 583 mm > 100 mm, So ok

Width of footing in z direction = 4000,

Allowable eccentricity = 4000/6 = 666 mm > 8.5 mm, So ok

Pressure on soil

$p = P/A \pm M_x / Z_x \pm M_y / Z_y$

p = Pressure on soil

P = Axial Compressive force

A = Area of footing

M_x, M_x = Bending moment

Z_x, Z_y = Section modulus

$p = 2700.3/(3.5 \times 4) \pm -271.85 \times 6/(3.5^2 \times 4) \pm -23 \times 6/(42 \times 3.5)$

p_{max} = 228.8 N/mm² < Allowable net bearing capacity of soil = 237.5 N/mm², So ok

p_{min} should be always positive to ensure full contact with soil. It is checked and found ok.

From above calculation, maximum soil pressure is 226.3 kN/m² < 237.5 kN/m², OK

Geometry of footing:

Considering total depth at end of footing = 400 mm = 0.4 m (D_{face})

Total depth of footing at column face = 750 mm = 0.75 m (D_{col})

So, effective depth at end of footing = 0.35 m (d_{face})

Effective depth at column face = 0.7 m (d_{col})

One-way shear check:

Critical section distance for one-way shear: as per clause no. 32.2.4.1a of IS 456, critical section is a vertical section from the face of the column/pedestal at a distance equal to the effective depth of footing (refer Figure 4.39).

Two-way shear check:

Two-way or Punching shear – As per Clause no. 31.6.1 of IS 456, critical section occurs at a distance equal to the half of effective depth of footing from the periphery of the column/ pedestal (refer Figure 4.39).

Detailing of Foundation:

The length of main bar of foundation from the vertical section of column face to end of the foundation before cover should not be less than development length in tension.

The length of embedment of column reinforcement in foundation from top face of footing should not be less than development length in tension.

The length of column reinforcement parallel to main bar of footing should not be less than 300 mm.

Minimum depth at outer face of trapezoidal footing should not be less than 150 mm. Reinforcement detailing as shown in Figure 4.45.

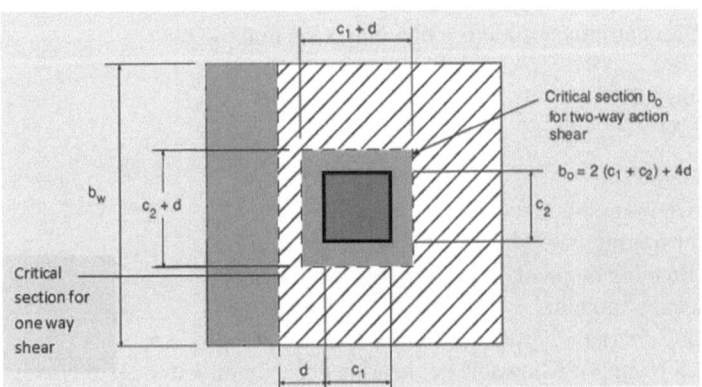

FIGURE 4.39 Critical sections of one-way and two-way shear stresses

Working drawings for different structural elements are given in Figures 4.39, 4.40, 4.41 4.42 4.43, 4.44.

4.4.15 Working Drawings of Slabs, Beams, Columns and Foundations

Reinforcement drawings in the form of working drawings for different structural elements, such as slabs, beams, columns and foundations, are available in Figures 4.40, 4.41, 4.42, 4.43, 4.44 and 4.45.

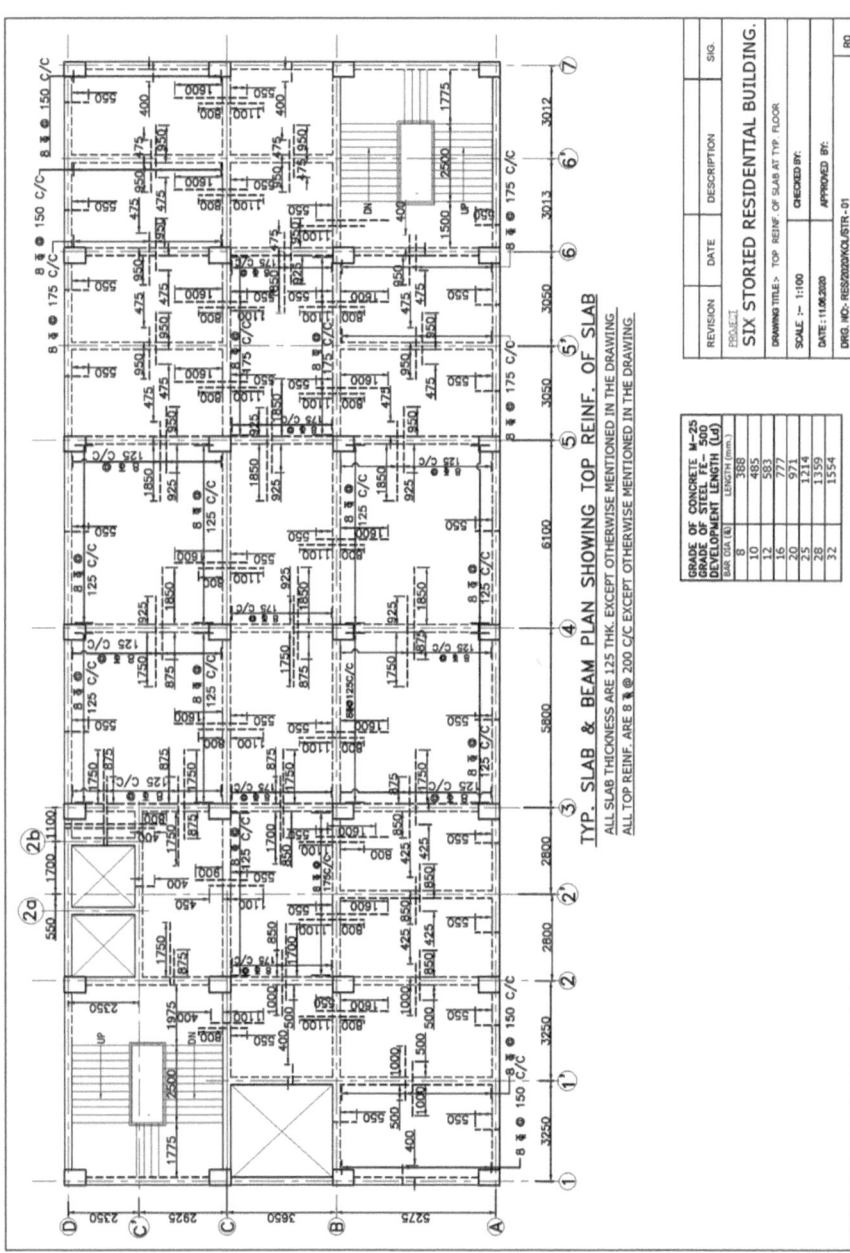

FIGURE 4.40 Detail of top reinforcements of slab

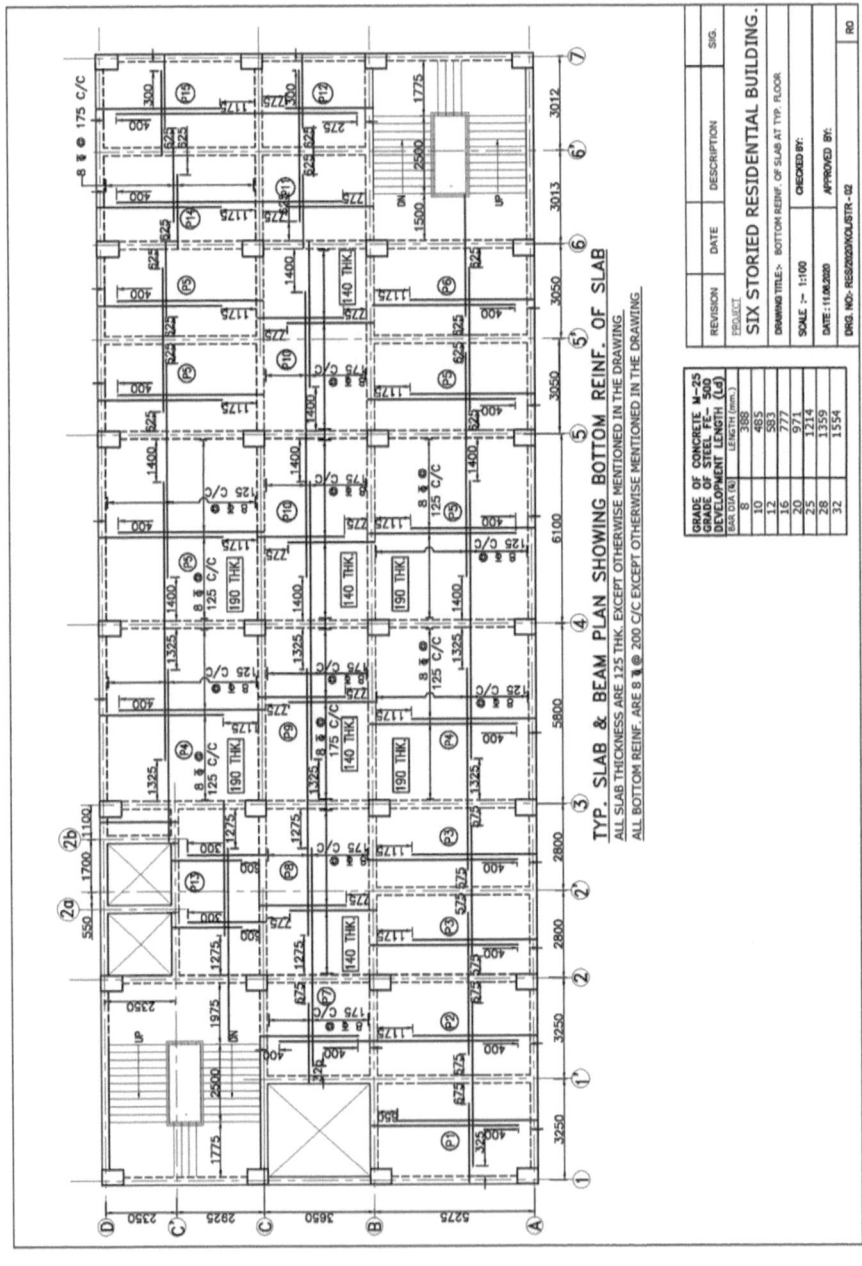

FIGURE 4.41 Detail of bottom reinforcements of slab

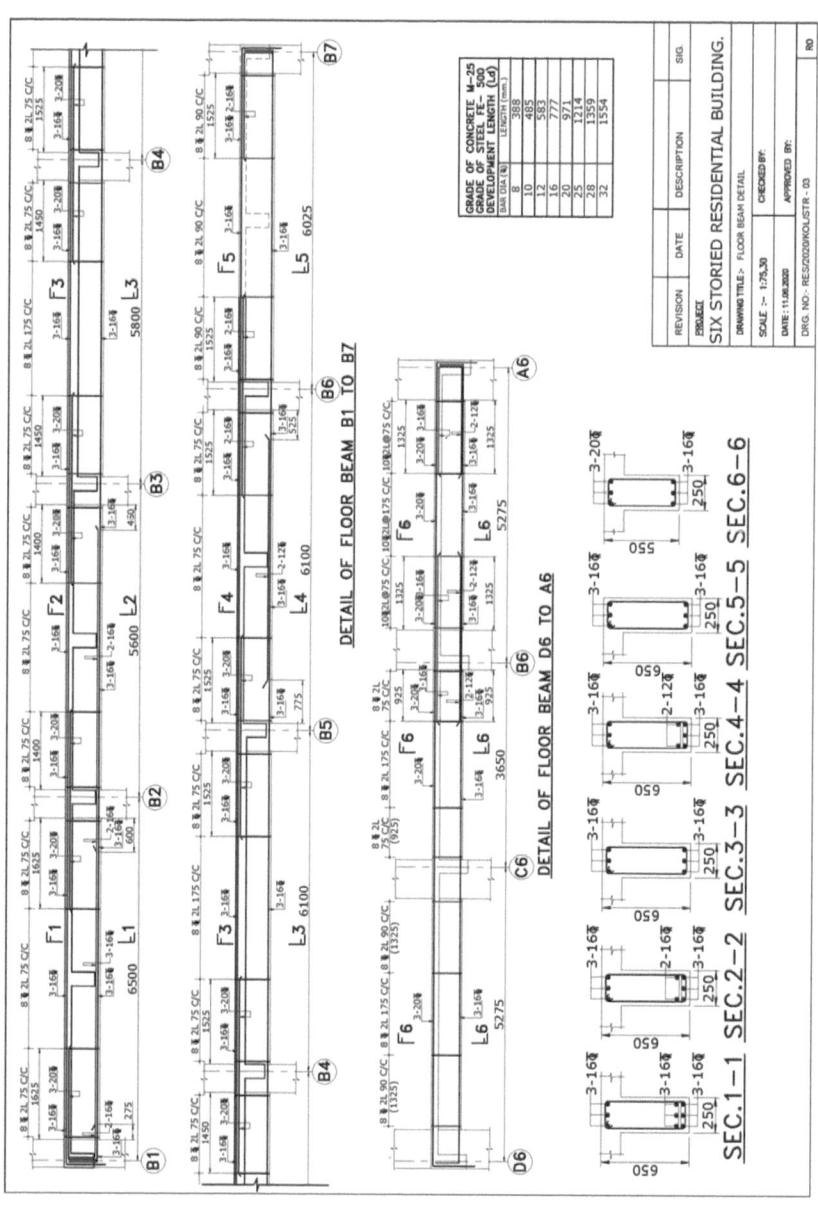

FIGURE 4.42 Working details of reinforcements for beams of the six-storied building

Note: Reinforcement chart of remaining beams may be represented in a similar manner.

TYP. FLOOR BEAM SCHEDULE

BEAM MKD	SIZE		BOTTOM REINFORCEMENT			TOP REINFORCEMENT			SHEAR STIRRUPS		
	B	D	LEFT	MID SPAN	RIGHT	LEFT	MID SPAN	RIGHT	LEFT	MID SPAN	RIGHT
D1-D2	250	650	3-16φ	3-16φ	3-16φ	3-16φ +2-16φ	3-16φ	3-16φ +3-20φ	8φ2L@90 C/C	8φ2L@175 C/C	8φ2L@90 C/C
D2-D3	250	650	3-16φ	3-16φ +2-12φ	3-16φ +2-12φ	3-16φ +3-20φ	3-16φ	3-16φ +3-20φ	10φ2L@75 C/C	10φ2L@75 C/C	10φ2L@75 C/C
D3-D4	250	650	3-16φ +2-12φ	3-16φ	3-16φ	3-16φ +3-20φ	3-16φ	3-16φ +2-16φ	8φ2L@75 C/C	8φ2L@175 C/C	8φ2L@75 C/C
D4-D5	250	650	3-16φ	3-16φ	3-16φ	3-16φ +2-16φ	3-16φ	3-16φ +3-20φ	8φ2L@90 C/C	8φ2L@175 C/C	8φ2L@90 C/C
D5-D6	250	650	3-16φ	3-16φ +2-12φ	3-16φ +2-12φ	3-16φ +3-20φ	3-16φ	3-16φ +3-20φ	8φ2L@75 C/C	8φ2L@75 C/C	8φ2L@75 C/C
D6-D7	250	650	3-16φ +2-12φ	3-16φ +2-12φ	3-16φ	3-16φ +3-20φ	3-16φ	3-16φ +3-20φ	8φ2L@75 C/C	8φ2L@75 C/C	8φ2L@75 C/C
C'2-C'3	250	650	3-16φ	3-16φ +2-12φ	3-16φ +2-12φ	3-12φ	3-12φ	3-12φ +3-12φ	8φ2L@75 C/C	8φ2L@75 C/C	8φ2L@75 C/C
C1-C2	250	650	3-16φ	3-16φ	3-16φ	3-16φ +3-20φ	3-16φ	3-16φ +3-20φ	8φ2L@90 C/C	8φ2L@125 C/C	8φ2L@90 C/C
C2-C3	250	650	3-16φ	3-16φ	3-16φ	3-16φ +3-20φ	3-16φ	3-16φ +2-16φ	8φ2L@90 C/C	8φ2L@175 C/C	8φ2L@90 C/C
C3-C4	250	650	3-16φ	3-16φ	3-16φ	3-16φ +2-16φ	3-16φ	3-16φ +2-16φ	8φ2L@90 C/C	8φ2L@175 C/C	8φ2L@90 C/C
C4-C5	250	650	3-16φ	3-16φ	3-16φ	3-16φ +2-16φ	3-16φ	3-16φ +2-16φ	8φ2L@90 C/C	8φ2L@175 C/C	8φ2L@90 C/C
C5-C6	250	650	3-16φ	3-16φ +2-12φ	3-16φ +2-12φ	3-16φ +2-16φ	3-16φ	3-16φ +3-20φ	8φ2L@75 C/C	8φ2L@75 C/C	8φ2L@75 C/C
C6-C7	250	650	3-16φ +2-16φ	3-16φ +3-16φ	3-16φ +2-16φ	3-16φ +3-20φ	3-16φ	3-16φ +3-20φ	8φ2L@75 C/C	8φ2L@75 C/C	8φ2L@75 C/C
B1-B2	250	650	3-16φ +2-16φ	3-16φ +3-16φ	3-16φ +2-16φ	3-16φ +3-20φ	3-16φ	3-16φ +3-20φ	8φ2L@75 C/C	8φ2L@75 C/C	8φ2L@75 C/C
B2-B3	250	650	3-16φ +2-16φ	3-16φ +2-16φ	3-16φ	3-16φ +3-20φ	3-16φ	3-16φ +3-20φ	8φ2L@75 C/C	8φ2L@75 C/C	8φ2L@75 C/C
B3-B4	250	650	3-16φ	3-16φ	3-16φ	3-16φ +3-20φ	3-16φ	3-16φ +3-20φ	8φ2L@75 C/C	8φ2L@175 C/C	8φ2L@75 C/C
B4-B5	250	650	3-16φ	3-16φ	3-16φ	3-16φ +3-20φ	3-16φ	3-16φ +3-20φ	8φ2L@75 C/C	8φ2L@175 C/C	8φ2L@75 C/C
B5-B6	250	650	3-16φ	3-16φ +2-12φ	3-16φ	3-16φ +3-20φ	3-16φ	3-16φ +2-16φ	8φ2L@75 C/C	8φ2L@75 C/C	8φ2L@75 C/C
B6-B7	250	650	3-16φ	3-16φ	3-16φ	3-16φ +2-16φ	3-16φ	3-16φ +2-16φ	8φ2L@90 C/C	8φ2L@90 C/C	8φ2L@90 C/C
A1-A2	250	650	3-16φ	3-16φ +2-12φ	3-16φ	3-16φ +3-20φ	3-16φ	3-16φ +3-20φ	8φ2L@75 C/C	8φ2L@75 C/C	8φ2L@75 C/C
A2-A3	250	650	3-16φ	3-16φ	3-16φ	3-16φ +3-20φ	3-16φ	3-16φ +3-20φ	8φ2L@75 C/C	8φ2L@75 C/C	8φ2L@75 C/C
A3-A4	250	650	3-16φ	3-16φ	3-16φ	3-16φ +3-20φ	3-16φ	3-16φ +2-16φ	8φ2L@90 C/C	8φ2L@175 C/C	8φ2L@90 C/C
A4-A5	250	650	3-16φ	3-16φ	3-16φ	3-16φ +2-16φ	3-16φ	3-16φ +3-20φ	8φ2L@90 C/C	8φ2L@175 C/C	8φ2L@90 C/C
A5-A6	250	650	3-16φ	3-16φ +2-12φ	3-16φ	3-16φ +3-20φ	3-16φ	3-16φ +3-20φ	8φ2L@75 C/C	8φ2L@75 C/C	8φ2L@75 C/C
A6-A7	250	650	3-16φ	3-16φ	3-16φ	3-16φ +3-20φ	3-16φ	3-16φ +2-16φ	8φ2L@90 C/C	8φ2L@175 C/C	8φ2L@90 C/C

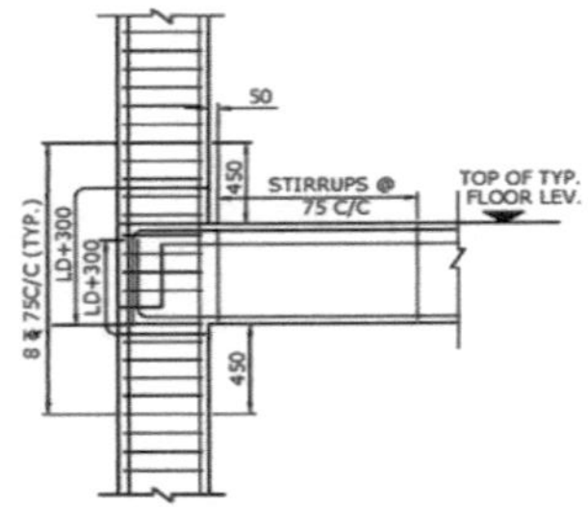

DETAIL OF BEAM COLUMN JUNCTION AT TYP. FLOOR LEV.

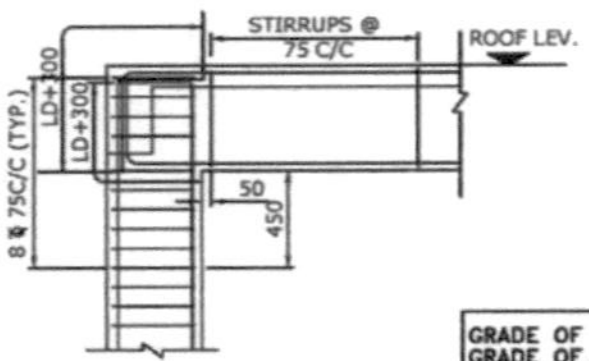

DETAIL OF BEAM COLUMN JUNCTION AT ROOF LEV.

GRADE OF CONCRETE M–25
GRADE OF STEEL FE– 500
DEVELOPMENT LENGTH (Ld)

BAR DIA (φ)	LENGTH (mm.)
8	388
10	485
12	583
16	777
20	971
25	1214
28	1359
32	1554

REVISION	DATE	DESCRIPTION	SIG.
PROJECT: SIX STORIED RESIDENTIAL BUILDING.			
DRAWING TITLE :- TYP. FLOOR BEAM SCHEDULE			
SCALE :- 1:100		CHECKED BY:	
DATE : 11.06.2020		APPROVED BY:	
DRG. NO:- RES/2020/KOL/STR - 04			R0

FIGURE 4.43 Schedule of reinforcements for beams of the six-storied building

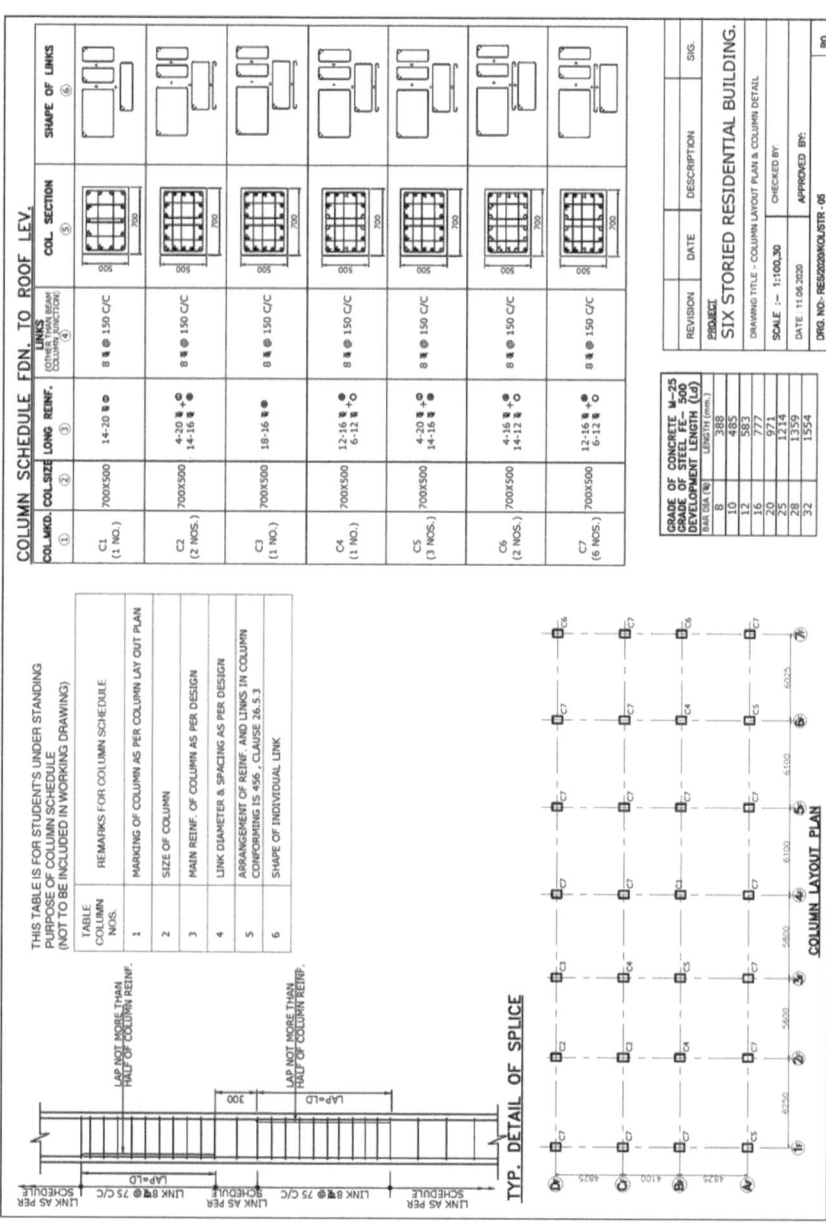

FIGURE 4.44 Working details of reinforcement for columns of the six-storied building

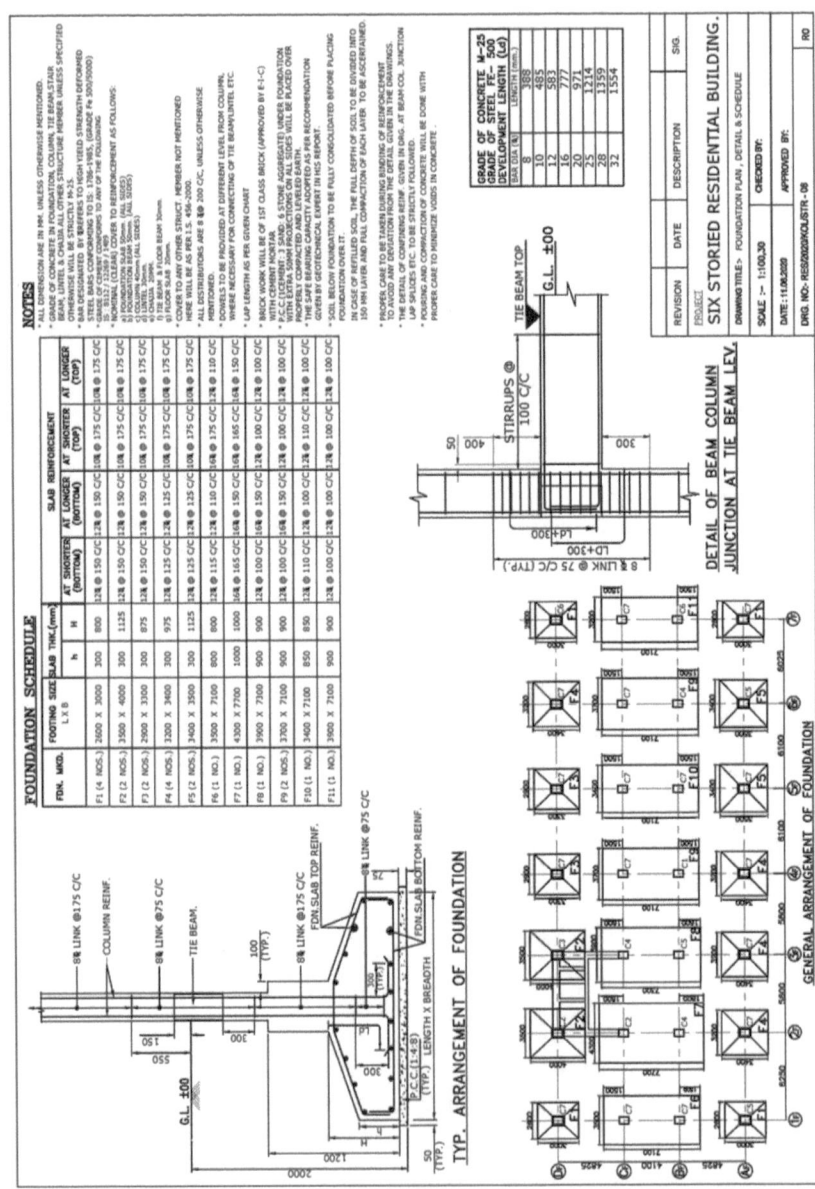

FIGURE 4.45 Working details of reinforcement for foundations of the six-storied building

4.5   DESIGN OF A 15-STORIED REINFORCED CONCRETE-FRAMED RESIDENTIAL BUILDING ON A PILE FOUNDATION

Basic Data

- Building plan as shown in Figure 4.46
- Building site: Kolkata, India
- Building on a pile foundation
- Grade of concrete M40 and grade of steel Fe500 TMT
- Typical floor to floor height = 3,100 mm
- Plinth height = 600 mm
- Foundation depth = 2,000 mm
- Walls other than partition walls: 250 mm-thick brickwork, including plaster
- Partition walls: 125 mm-thick brickwork, including plaster
- Floor finish: 9.8 mm-thick vitrified tiles over 25 mm-thick plain concrete
- Ceiling plaster: 12 mm-thick cement plaster

4.5.1   DEAD LOAD AND LIVE LOADS

Dead Load

As per IS 875 (Part 1), 2015

Unit weight of reinforced cement concrete = 25 kN/m³
Unit weight of plain cement concrete = 24 kN/m³
Unit weight of ceiling plaster = 20.4 kN/m³
Unit weight of brick wall with plaster = 18.85 kN/m³

For a typical floor

Self-weight of floor slab (125 mm thick) = (0.125) × (25) =   3.125 kN/m²
Self-weight of 25 mm-thick concrete below tiles = (0.025) × (24) =   0.6 kN/m²
Self-weight of 9.8 mm-thick vitrified tiles (from company handbook) =   0.185 kN/m2
Self-weight of 12 mm ceiling plaster = (0.012) × (20.4) =   0.24 kN/m²

≈ 4.2 kN/m²

For roof

Self-weight of roof slab (125 mm thick) = (0.125) × (25) =   3.125 kN/m²
Self-weight of concrete waterproofing layer = (0.125) × (24) =   3.0 kN/m²
Self-weight of 12 mm ceiling plaster = (0.012) × (20.4) =   0.24 kN/m²

≈ 6.4 kN/m²

Brick walls at ground floor level

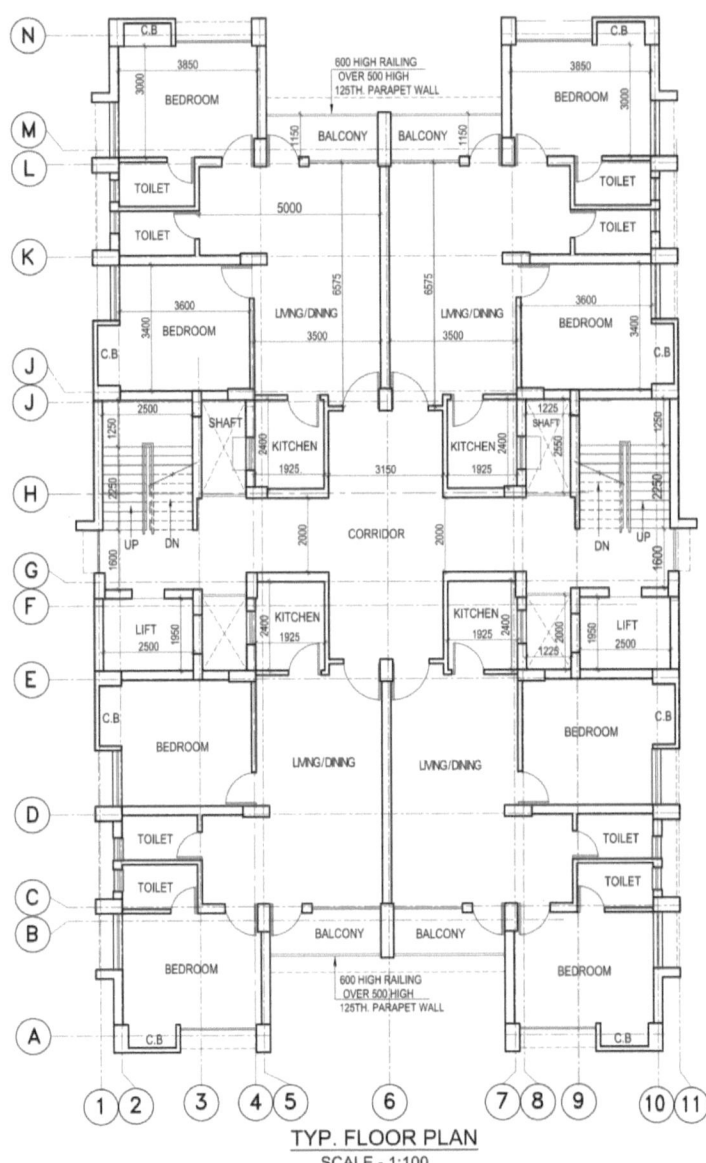

TYP. FLOOR PLAN
SCALE - 1:100

FIGURE 4.46 Building plan

Depth of floor beam rib assumed initially = 0.5 m

Height of brick walls from top of tie beam to bottom of first floor beam = (floor-to-floor height + plinth height –depth of first floor beam rib) = 3.1 + 0.6 – 0.5 = 3.2 m

Self-weight of 3.2 m-high, 250 mm-thick brick walls (including plaster)

$$= (3.2) \times (0.25) \times (18.85) = 15.2 \text{ kN/m}$$

Self-weight of 3.2 m-high, 125 mm-thick brick walls (including plaster)

$$= (3.2) \times (0.125) \times (18.85) = 7.6 \text{ kN/m}$$

Brick walls at typical floor level

Height of brick walls (floor to floor height – depth of floor beam rib)

$$= 3.1 - 0.5 = 2.6 \text{ m}$$

Self-weight of 2.6 m-high, 250 mm-thick brick walls (including plaster)

$$= (2.6) \times (0.25) \times (18.85) = 12.35 \text{ kN/m}$$

Self-weight of 2.6 m-high, 125 mm-thick brick walls (including plaster)

$$= (2.6) \times (0.125) \times (18.85) = 6.18 \text{ kN/m}$$

Self-weight of 1 m-high, 250 mm-thick brick walls (including plaster)

$$= (1) \times (0.25) \times (18.85) = 4.75 \text{ kN/m}$$

Imposed load/live load as per IS 875 (Part 2), 2015

For rooms	2	kN/m^2
For corridor and stairs	3	kN/m^2
For toilet	2	kN/m^2

4.5.2 WIND ANALYSIS

4.5.2.1 Basic Wind Pressure

As per IS 875 (Part 3), 2015

Building/site location: Kolkata, India

V_b = basic wind speed, as per figure 1 of IS 875 (Part 3), 2015 = 50 m/s

As per clause 6.3 of IS 875 (Part 3), 2015

V_z = design wind speed at a height z (m) from ground level in m/s

$$= (V_b) (k_1) (k_2) (k_3) (k_4)$$

As per clause 6.3.1 and Table 1 of IS 875 (Part 3), 2015, the risk coefficient k_1 = 1.0 for all general buildings and structures for basic wind speed of 50 m/s.

The building site is regarded as a city area, where there are numerous closely spaced buildings around 10 m high with a few tall buildings as well. As per clause 6.3.2.1, it falls into terrain category 3.

As per clause 6.3.2.2 and Table 2 of IS 875 (Part 3), 2015, the k_2 factors for different heights are calculated and furnished in Table 4.80.

TABLE 4.80
Terrain factor (k_2) at different heights

Height (m)	k_2
10	0.91
15	0.97
20	1.01
30	1.06
50	1.12

As per clause 6.3.3.1 of IS 875 (Part 3), 2015, the topography factor k_3 = 1.0 is taken, because the building site is more or less flat – i.e., the upwind slope is less than 3^0. It may be noted here that, for upwind slopes of more than 3^0, the value of k_3 has to be calculated as per the recommendations given in annexure C of IS 875 (Part 3), 2015. As Kolkata is not within a cyclonic region, k_4 is not applicable. The design wind speeds at different heights are calculated and presented in Table 4.81.

TABLE 4.81
Design wind speed at different heights

Height (m)	V_b (m/s)	k_1	k_2	k_3	V_z (m/s)
10	50	1	0.91	1	45.5
15	50	1	0.97	1	48.5
20	50	1	1.01	1	50.5
30	50	1	1.06	1	53.0
50	50	1	1.12	1	56.0

As per clause 7.2 of IS 875 (Part 3), 2015, the wind pressure at height z (m):

$$p_z = 0.6V_z^2$$

TABLE 4.82
Wind pressure at different heights

Height (m)	V_z (m/s)	p_z (N/m²)
10	45.5	1,242
15	48.5	1,411
20	50.5	1,530
30	53.0	1,685
50	56.0	1,882

The wind pressure at different heights has been computed in Table 4.82.
As per clause 7.2.1 of IS 875 (Part 3), 2015, the design wind pressure:

$$p_d = K_d \cdot K_a \cdot K_c \, p_z$$

However, p_d should not exceed $0.7p_z$.
As per clause 7.2.1 of IS 875 (Part 3), 2015:

Wind directionality factor $(K_d) = 0.9$
Length of building $= 26.45$ m
Width of building $= 15.95$ m
Height of building $= 49.7$ m
Tributary area in both longer and shorter directions $= (26.45 \text{ m} \times 15.95 \text{ m}) > 100 \text{ m}^2$

As per clause 7.2.2 and Table 4 of IS 875 (Part 3), 2015, the area averaging factor $k_a = 0.8$
As per clause 7.3.13 of IS 875 (Part 3), 2015, the combination factor $k_c = 0.9$
The design wind pressure at different heights has been computed in Table 4.83.

TABLE 4.83
Design wind pressure

Height (m)	p_z (N/m²)	k_d	k_a	k_c	p_d	$0.7p_z$	Calculated p_d (N/m²)
10	1,242	0.9	0.8	0.9	805	869	869
15	1,411	0.9	0.8	0.9	914	945	945
20	1,530	0.9	0.8	0.9	991	1,011	1,011
30	1,685	0.9	0.8	0.9	1,092	1,180	1,180
50	1,882	0.9	0.8	0.9	1,219	1,317	1,317

4.5.2.2 Wind Load as per "Drag Coefficient Approach"

All as per IS 875 (Part 3), 2015.

As per clause 7.1, the wind load on a building may be calculated by considering it as a closed building as a whole, like a vertical cantilever

As per clause 7.4, the wind load on building may be calculated as follows:

$$F = C_f \cdot A_e \cdot P_d$$

where

F = force acting in a specified direction on the frontal area

C_f = force coefficient of the building, as per clause 7.4.2

A_e = effective frontal area of the building

P_d = design wind pressure

The force coefficients for differently shaped buildings are given in figures 4 and 5 and Table 25 of IS 875 (Part 3), 2015.

C_f is calculated as per clause 7.4.2.1 and Figure 4 of IS 875 (Part 3), 2015, as follows.

Wind along X direction:

width of the building (a) = 15.95 m
length of the building (b) = 26.45 m
height of the building (h) = 49.7 m

$$a/b = 0.6, \ h/b = 1.9$$

As per clause 7.4.2.1 and Figure 4 of IS 875 (Part 3), 2015, C_f = 1.25.

Wind along Z direction:

width of the building (a) = 26.45 m
length of the building (b) = 15.95 m
height of the building (h) = 49.7 m

$$a/b = 1.65, \ h/b = 3$$

As per clause 7.4.2.1 and figure 4 of IS 875 (Part 3), 2015, C_f = 1.2.

As an input to STAAD Pro CE, wind is defined in terms of its load intensity (in kN/m²) at different heights; these data are obtained, and furnished below, and the corresponding effective frontal area is calculated by the software itself.

The design wind force (kN/m²) at different heights in both principal horizontal directions, X and Z, have been computed in Tables 4.84 and 4.85.

TABLE 4.84
Wind force along the X direction

Height (m)	C_f	p_d (kN/m²)	Wind force (kN/m²)
10	1.25	0.869	1.09
15	1.25	0.945	1.81
20	1.25	1.011	1.26
30	1.25	1.180	1.48
50	1.25	1.317	1.65

TABLE 4.85
Wind force along the Z direction

Height (m)	C_f	p_d (kN/m²)	Wind force (kN/m²)
10	1.2	0.869	1.04
15	1.2	0.945	1.13
20	1.2	1.011	1.21
30	1.2	1.180	1.42
50	1.2	1.317	1.58

4.5.2.3 Wind Load as per "Pressure Coefficient Method"

As per clause 7.1 of IS 875 (Part 3), 2015, wind loads on buildings also have to be calculated as follows.

The wind load can be calculated on individual structural elements – i.e., roof and wall elements, etc. – using the internal pressure coefficient (C_{pi}) and the external pressure coefficient (C_{pe}).

The pressure coefficient, as per clause 7.3 of IS 875 (Part 3), 2015, states that the pressure coefficients are always given for a particular surface or part of the surface of a building. The wind load acting normal to a surface is obtained by multiplying the area of that surface or its appropriate portion by the pressure coefficient (C_p) and the design wind pressure at the height of the surface from the ground.

As per clause 7.3.1 of IS 875 (Part 3), 2015:

$$F = (C_{pe} - C_{pi}) A.p_d$$

where

C_{pe} = external pressure coefficient
C_{pi} = internal pressure coefficient
A = surface area of structural element (i.e., wall or roof)
p_d = design wind pressure
Width of the building (w) = 15.95 m
Length of the building (l) = 26.45 m

Height of the building (h) = 49.7 m

l/w = 1.66, h/w = 2.99

As per table 5 of IS 875 (Part 3), 2015, the external pressure coefficients for rectangular buildings are calculated and furnished in Table 4.86. The wind angle and marked external surfaces are shown in Figure 4.47.

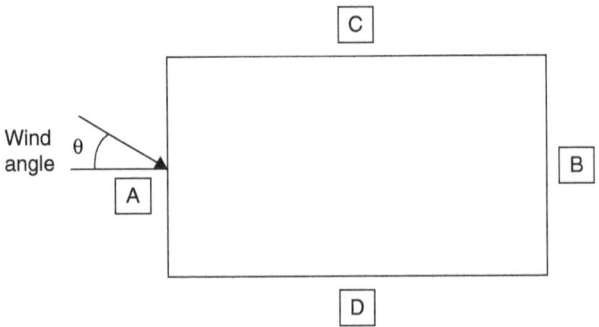

FIGURE 4.47   Plan showing wind angle and marking of external wall faces

TABLE 4.86
External pressure coefficients

C_{pe} for surface (wind angle 0°)

A	B	C	D
0.7	−0.4	−0.7	−0.7

When wind angle 90°C_{pe} for surface

A	B	C	D
−0.5	0.5	0.8	−0.1

A positive sign indicates that the wind force is acting toward the surface (compression).

A negative sign indicates that the wind force is acting away from the surface (suction).

Percentage of openings on external surface = total area of openings × 100 / solid area

In terms of buildings with a medium amount of openings (5 to 20 percent of the wall area), as per clause 7.3.2.2, the internal pressure coefficient C_{pi} = ±0.5.

The combined effects of the external pressure coefficient and the internal pressure coefficient are furnished in Tables 4.86 and 4.87.

Taking into account the different pressure coefficients for different external surfaces, wind load intensities are obtained as per Table 4.88 at different levels along the height of the building.

TABLE 4.87
Net pressure coefficients considering external and
internal pressure coefficients

	C_{pe} for surface (wind angle 0°)			
	A	B	C	D
	0.7	−0.4	−0.7	−0.7
	C_{pi} for medium openings			
Case 1	0.5	0.5	0.5	0.5
Case 2	−0.5	−0.5	−0.5	−0.5
$C_{pe} - C_{pi}$ for case1	0.2	−0.9	−1.2	−1.2
$C_{pe} - C_{pi}$ for case2	1.2	0.1	−0.2	−0.2
	C_{pe} for surface (wind angle 90°)			
	A	B	C	D
	−0.5	−0.5	0.8	−0.1
	C_{pi} for medium openings			
Case 1	0.5	0.5	0.5	0.5
Case 2	−0.5	−0.5	−0.5	−0.5
$C_{pe} - C_{pi}$ for case1	−1.0	−1.0	0.3	−0.6
$C_{pe} - C_{pi}$ for case2	0	0	1.3	0.4

TABLE 4.88
Wind force calculation

Wind angle 0°

		Wind force on different surfaces (kN/m²)			
Height (m)	p_d (N/m²)	A	B	C	D
Case 1					
10	869	0.17	-0.7821	-1.0428	-1.0428
15	945	0.19	-0.8505	-1.134	-1.134
20	1,011	0.2	-0.9099	-1.2132	-1.2132
30	1,180	0.24	-1.062	-1.416	-1.416
50	1,317	0.26	-1.1853	-1.5804	-1.5804
Case 2					
10	869	1.04	0.0869	-0.1738	-0.1738
15	945	1.13	0.0945	-0.189	-0.189
20	1,011	1.21	0.1011	-0.2022	-0.2022
30	1,180	1.42	0.118	-0.236	-0.236
50	1,317	1.58	0.1317	-0.2634	-0.2634

Wind angle 90°

Height (m)	p_d (N/m²)	Wind force on different surface (kN/m²)			
		A	B	C	D
Case 1					
10	869	-0.87	-0.869	0.2607	-0.5214
15	945	-0.95	-0.945	0.2835	-0.567
20	1,011	-1.01	-1.011	0.3033	-0.6066
30	1,180	-1.18	-1.18	0.354	-0.708
50	1,317	-1.32	-1.317	0.3951	-0.7902
Case 2					
10	869	0	0	1.1297	0.3476
15	945	0	0	1.2285	0.378
20	1,011	0	0	1.3143	0.4044
30	1,180	0	0	1.534	0.472
50	1,317	0	0	1.7121	0.5268

4.5.2.4 Wind Load as per "Gust Factor Approach"

As per clause 9.1 of IS 875 (Part 3), 2015:

length of the building = 26.45 m
width of the building = 15.95 m
height of the building = 49.7 m
H/b = 1.87 < 5

The natural frequency of the building in the first mode is calculated using STAAD Pro CE software

$$= 0.347 \text{ Hz} < 1\text{Hz}$$

Therefore, wind-induced dynamic effects have to be examined. The natural frequencies from the STAAD Pro CE analysis (calculated frequencies for the dead and live loads) are shown in Table 4.89.

TABLE 4.89
Natural frequencies

Mode	Frequency (cycles/sec)	Period (sec.)
1	0.347	2.87841
2	0.377	2.65019
3	0.383	2.61158
4	1.163	0.86009
5	1.186	0.84350
6	1.204	0.83029

Vortex Shedding

As per clause 9.2.1, note 4, of IS 875 (Part 3), 2015.

L = longer side of the building = 26.45 m
B = shorter side of the building = 15.95 m

$$\textbf{L/B = 1.658 < 2}$$

No vortex shedding effect therefore needs to be considered.

Along-Wind Effect

As per clause 9.2.1, note 4, of IS 875 (Part 3), 2015:

$$\textbf{F}_z = \textbf{C}_{f,z}\textbf{A}_z\textbf{p}_d\textbf{G}$$

where

F_z = design peak along-wind load on the building structure at any height z
$C_{f,z}$ = drag force coefficient of the building structure corresponding to the area A_z
p_d = design hourly mean wind pressure
= $\textbf{0.6 V}_{zd}^2$ (in N/m²)
V_{zd} = design hourly mean wind speed at height z (in m/s)
A_z = effective frontal area of the building structure at any height z (in m²)
G = gust factor = $\textbf{1 + r } \sqrt{[\textbf{g}_v^2\textbf{B}_s(\textbf{1 + Ø})^2 + \textbf{H}_s\textbf{g}_r^2\textbf{SE/ß}]}$

where

r = roughness factor, which is twice the longitudinal turbulence intensity
g_v = peak factor for upwind velocity fluctuation (3.0 for category 1 and 2 terrains, and 4.0 for category 3 and 4 terrains)
B_s = background factor, indicating the measure of slowly varying components of fluctuating wind load caused by lower-frequency wind speed variations

$$\textbf{B}_s = \textbf{1 / [1 + } \sqrt{(\textbf{0.26(h–s)}^2 + \textbf{0.46 b}_{sh}^2/\textbf{L}_h)}]$$

where

b_{sh} = average breadth of the building/structure between heights s and h
L_h = measure of effective turbulence length scale at height h (in m)

$$= \textbf{85 [h/10]}^{0.25} \text{ for terrain categories 1 to 3}$$
$$= \textbf{70 [h/10]}^{0.25} \text{ for terrain category 4}$$

Ø = factor to account for the second-order turbulence intensity = $\textbf{g}_v\textbf{I}_{h,I} \sqrt{\textbf{B}_s}/\textbf{2}$

where

$I_{h,i}$ = turbulence intensity at height h in terrain category 1
H_s = height factor for resonance response = $1 + (s/h)^2$
S = size reduction factor

$$= 1/[1 + (3.5f_a h)/V_{hd}][1 + (4f_a b_{oh})/V_{hd}]$$

where

b_{oh} = average breadth of the building/structure between 0 and h
E = spectrum of turbulence in the approaching wind stream

$$= A N/ (1 + 70.8 N^2)^{5/6}$$

where

N = Effective reduced frequency

$$= f_a L_h/V_{hd}$$

where

f_a = first-mode natural frequency of the building/structure in along-wind direction (in Hz)
V_{hd} = design hourly mean wind speed at height h (in m/s)
ß = damping coefficient of the building/structure (for Reinforced Concrete structures, ß = 0.020)
g_r = peak factor for resonant response = $\sqrt{2\ln(3600f_a)}$

Calculation of Gust Factor (G)

Calculation of p_d and V_{zd}.
  We know

V_b = 50 m/s
k_1 = 1.0
k_3 = 1.0
k_4 = not applicable in this case, as the site is not in a coastal area

As per clause 6.4 of IS 875 (Part 3), 2015:

Hourly mean wind speed:

$$V_{zH} = k_{2i}V_b$$

where

k_{2i} = hourly mean wind speed factor for terrain category 3 (the category for this building)

$$= 0.1423[\ln(z/z_{0i})] \, (z_{0i})^{0.0706}$$

where

z = height or distance above the ground
z_{0i} = aerodynamic roughness height for ith terrain
Design hourly mean wind speed:

$$V_{zd} = V_b \, k_1 \, k_{2i} \, k_3 \, k_4$$

Typical calculation for k_{2i}, V_{zd} and p_d for z = 3.7 m, as per clause 6.3.2.1c of IS 875 (Part 3), 2015

Aerodynamic roughness height for terrain category 3 = z_{03} = 0.2
z = 3.7 m
k_{2i} = 0.1423[ln(3.7/0.2)](0.2)$^{0.0706}$ = 0.370603
V_{zd} = (50)(1)(0.370603)(1) = 18.53 m/s
p_d = (0.6) $(V_{zd})^2$/1000 = (0.6)(18.53)^2/1000 = 0.21 kN/m^2

In a similar way, all these parameters are obtained for different levels (z), and furnished in Table 4.90.

TABLE 4.90
Calculation of p_d

z (m)	z_{03}	k_{2i}	V_{zd}	p_d (kN/m^2)
3.7	0.2	0.37	18.5	0.21
6.8	0.2	0.45	22.5	0.30
9.9	0.2	0.5	25.0	0.38
13	0.2	0.53	26.5	0.42
16.1	0.2	0.56	28.0	0.47
19.2	0.2	0.58	29.0	0.50
22.3	0.2	0.60	30.0	0.54
25.4	0.2	0.62	31.0	0.58
28.5	0.2	0.63	31.5	0.60
31.6	0.2	0.64	32.0	0.61
34.7	0.2	0.65	32.5	0.63
37.8	0.2	0.67	33.5	0.67
40.9	0.2	0.68	34.0	0.69
44	0.2	0.69	34.5	0.71
47.1	0.2	0.69	34.5	0.71
49.7	0.2	0.70	35.0	0.74

Note: z (m) = height of floor levels.

Turbulence Intensity (I_{hi})

As per clause 6.5 of IS 875 (Part 3), 2015

Terrain category 1 – $I_{z,1} = 0.3507 - 0.0535 \log_{10}(z/z_{01})$
Terrain category 2 – $I_{z,2} = I_{z,1} + 1/7(I_{z,4} - I_{z,1})$
Terrain category 3 – $I_{z,1} + 3/7(I_{z,4} - I_{z,1})$
Terrain category 4 – $I_{z,4} = 0.466 - 0.1358 \log_{10}(z/z_{0,4})$
Turbulence intensity for terrain category 1 = $I_{z,1} = 0.3507 - 0.0535 \log_{10}(z/z_{01})$
Taking z = 3.7 m, from clause 6.3.2.1, $z_{01} = 0.002$ m, $z_{04} = 2$ m
I_{hi} for terrain category 1 = $0.3507 - 0.0535 \log_{10}(3.7/0.002) = 0.1759$
I_{hi} for terrain category 4 = $0.466 - 0.1358 \log_{10}(3.7/2) = 0.4297$
I_{hi} for terrain category 3 = $0.1759 + (3/7)(0.4297 - 0.1759) = 0.2847$
r = roughness factor for category 3 = $2 I_{h3} = (2)(0.2847) = 0.5694$

In a similar way, all these parameters are obtained for different levels (z), and furnished in Table 4.91.

TABLE 4.91
Calculation of turbulence intensity and roughness factor

z (m)	z_{01}	z_{04}	I_{hi}Cat 1	I_{hi}Cat 4	I_{hi}Cat 3	r for category 3
3.7	0.002	2	0.1759	0.4297	0.2847	0.5694
6.8	0.002	2	0.1617	0.3938	0.2612	0.5224
9.9	0.002	2	0.153	0.3717	0.2467	0.4934
13	0.002	2	0.1467	0.3556	0.2362	0.4724
16.1	0.002	2	0.1417	0.343	0.228	0.456
19.2	0.002	2	0.1376	0.3326	0.2212	0.4424
22.3	0.002	2	0.1341	0.3238	0.2154	0.4308
25.4	0.002	2	0.1311	0.3161	0.2104	0.4208
28.5	0.002	2	0.1284	0.3093	0.2059	0.4118
31.6	0.002	2	0.126	0.3032	0.2019	0.4038
34.7	0.002	2	0.1239	0.2977	0.1984	0.3968
37.8	0.002	2	0.1219	0.2927	0.1951	0.3902
40.9	0.002	2	0.12	0.288	0.192	0.384
44	0.002	2	0.1183	0.2837	0.1892	0.3784
47.1	0.002	2	0.1168	0.2797	0.1866	0.3732
49.7	0.002	2	0.1155	0.2765	0.1845	0.369

Background Factor (B_s) and Height Factor for Resonance Response (H_s)

A typical calculation proceeds as follows.

Taking s = level on a building/structure for evaluation of the along-wind load effects = 3.7 m
h = height of the structure above mean ground level = 49.7 m

L_h = measure of effective turbulence length scale at height h (for terrain category 3)
= 85 [49.7/10]$^{0.25}$ = 126.91
b_{sh} = average breadth of the building/structure between heights s = 16.701 m
B_s = background factor = $1/[1 + \{\sqrt{((0.26)(49.7 - 3.7)^2 + 0.46(16.701)^2)}/126.91\}] = 0.8297$
H_s = height factor for resonance response = $1 + (3.7/49.7)^2 = 1.0055$

In a similar way, all these parameters are obtained for different levels (z), and furnished in Table 4.92.

TABLE 4.92
Effective turbulence length, background factor, height factor, etc.

h	s	L_h for category 3	bsh	Bs	Hs
49.7	3.7	126.91	16.701	0.8297	1.0055
49.7	6.8	126.91	16.701	0.8374	1.0187
49.7	9.9	126.91	16.701	0.8452	1.0397
49.7	13	126.91	16.701	0.853	1.0684
49.7	16.1	126.91	16.701	0.8607	1.1049
49.7	19.2	126.91	16.701	0.8684	1.1492
49.7	22.3	126.91	16.701	0.8759	1.2013
49.7	25.4	126.91	16.701	0.8832	1.2612
49.7	28.5	126.91	16.701	0.8902	1.3288
49.7	31.6	126.91	16.701	0.8968	1.4043
49.7	34.7	126.91	16.701	0.9028	1.4875
49.7	37.8	126.91	16.701	0.9081	1.5785
49.7	40.9	126.91	16.701	0.9124	1.6772
49.7	44	126.91	16.701	0.9156	1.7838
49.7	47.1	126.91	16.701	0.9175	1.8981
49.7	49.7	126.91	16.701	0.9181	2

$\emptyset$ = Factor to account for the second-order turbulence intensity = $g_v I_{h,i} \sqrt{B_s}/2$
Typical calculation for z = 3.7 m
g_v = Peak factor for upwind velocity fluctuation for terrain category 3 = 4
$\emptyset$ = $(4)(0.2847)\sqrt{(0.8297)}/2 = 0.5187$

In a similar way, all these parameters are obtained for different levels (z) and furnished in Table 4.93.

f_a = first mode natural frequency of the building/structure in the along-wind direction
= 0.347 Hz (already obtained earlier using STAAD Pro CE software)
b_{oh} = average breadth of the building/structure between 0 and h = 16.7 m
S = $1/[1 + ((3.5)(0.286)(49.7)/18.5)][1 + ((4)(0.286)(16.7)/18.5] = 0.1334$

In a similar way, all these parameters are obtained for different levels (z) and furnished in Table 4.94.

TABLE 4.93
Peak factor, factor to account for the second-order turbulence intensity

z	g_v	I_{hi}	Bs	ø
3.7	4	0.2847	0.8297	0.5187
6.8	4	0.2612	0.8374	0.478
9.9	4	0.2467	0.8452	0.4536
13	4	0.2362	0.853	0.4363
16.1	4	0.228	0.8607	0.423
19.2	4	0.2212	0.8684	0.4123
22.3	4	0.2154	0.8759	0.4032
25.4	4	0.2104	0.8832	0.3955
28.5	4	0.2059	0.8902	0.3885
31.6	4	0.2019	0.8968	0.3824
34.7	4	0.1984	0.9028	0.377
37.8	4	0.1951	0.9081	0.3718
40.9	4	0.192	0.9124	0.3668
44	4	0.1892	0.9156	0.3621
47.1	4	0.1866	0.9175	0.3575
49.7	4	0.1845	0.9181	0.3536

TABLE 4.94
Size reduction factor

z	f_a	h	b_{oh}	V_{hd}	S
3.7	0.347	49.7	16.7	18.5	0.1041
6.8	0.347	49.7	16.7	22.5	0.1338
9.9	0.347	49.7	16.7	25	0.152
13.0	0.347	49.7	16.7	26.5	0.1627
16.1	0.347	49.7	16.7	28	0.1734
19.2	0.347	49.7	16.7	29	0.1804
22.3	0.347	49.7	16.7	30	0.1873
25.4	0.347	49.7	16.7	31	0.1941
28.5	0.347	49.7	16.7	31.5	0.1975
31.6	0.347	49.7	16.7	32	0.2009
34.7	0.347	49.7	16.7	32.5	0.2043
37.8	0.347	49.7	16.7	33.5	0.2109
40.9	0.347	49.7	16.7	34	0.2143
44.0	0.347	49.7	16.7	34.5	0.2175
47.1	0.347	49.7	16.7	34.5	0.2175
49.7	0.347	49.7	16.7	35	0.2208

Note: Typical calculation of size reduction factor (S) for z = 3.7 m.

Typical calculation of spectrum of turbulence in the approaching wind stream (E) and effective reduced frequency (N), for z = 3.7 m:

N = (0.347)(126.91)/18.5 = 2.3804
E = (22/7)(2.3804)/(1 + (70.8)(2.3804)^2)$^{5/6}$ = 0.0505

In a similar way, all these parameters are obtained for different levels (z) and furnished in Table 4.95.

TABLE 4.95
Spectrum turbulence and effective reduced frequency

z	f_a	L_h	V_{hd}	N	E
3.7	0.347	126.91	18.5	2.3804	0.0505
6.8	0.347	126.91	22.5	1.9572	0.0575
9.9	0.347	126.91	25	1.7615	0.0616
13	0.347	126.91	26.5	1.6618	0.064
16.1	0.347	126.91	28	1.5728	0.0664
19.2	0.347	126.91	29	1.5185	0.068
22.3	0.347	126.91	30	1.4679	0.0695
25.4	0.347	126.91	31	1.4206	0.071
28.5	0.347	126.91	31.5	1.398	0.0717
31.6	0.347	126.91	32	1.3762	0.0725
34.7	0.347	126.91	32.5	1.355	0.0732
37.8	0.347	126.91	33.5	1.3146	0.0747
40.9	0.347	126.91	34	1.2952	0.0754
44	0.347	126.91	34.5	1.2765	0.0761
47.1	0.347	126.91	34.5	1.2765	0.0761
49.7	0.347	126.91	35	1.2582	0.0769

Typical calculation of gust factor (G) for z = 3.7 m:

g_r = √2ln(3,600f_a) = √2ln((3,600)(0.286)) = 3.7248
ß = damping coefficient of the building/structure; for an Reinforced Concrete structure, ß = 0.020
G = 1 + r √[$g_v^2 B_s$(1 + ø)^2 + $H_s g_r^2$SE/ß]
= 1 + (0.5694)√[(4)^2(0.8297)(1 + 0.5187)^2 + ((1.0055)(3.7248)^2(0.1041)
(0.0505)/0.02)]
= 4.33

In a similar way, all these parameters are obtained for different levels (z) and furnished in Table 4.96.

TABLE 4.96
Gust factor (G)

Height (m)	g_v	B_s	ø	H_s	g_r	S	E	ß	r	G
3.7	4	0.8297	0.5187	1.0055	3.7248	0.1041	0.0505	0.02	0.5694	4.33
6.8	4	0.8374	0.478	1.0187	3.7248	0.1338	0.0575	0.02	0.5224	4.08
9.9	4	0.8452	0.4536	1.0397	3.7248	0.152	0.0616	0.02	0.4934	3.93
13	4	0.853	0.4363	1.0684	3.7248	0.1627	0.064	0.02	0.4724	3.83
16.1	4	0.8607	0.423	1.1049	3.7248	0.1734	0.0664	0.02	0.456	3.76
19.2	4	0.8684	0.4123	1.1492	3.7248	0.1804	0.068	0.02	0.4424	3.71
22.3	4	0.8759	0.4032	1.2013	3.7248	0.1873	0.0695	0.02	0.4308	3.67
25.4	4	0.8832	0.3955	1.2612	3.7248	0.1941	0.071	0.02	0.4208	3.65
28.5	4	0.8902	0.3885	1.3288	3.7248	0.1975	0.0717	0.02	0.4118	3.62
31.6	4	0.8968	0.3824	1.4043	3.7248	0.2009	0.0725	0.02	0.4038	3.6
34.7	4	0.9028	0.377	1.4875	3.7248	0.2043	0.0732	0.02	0.3968	3.6
37.8	4	0.9081	0.3718	1.5785	3.7248	0.2109	0.0747	0.02	0.3902	3.61
40.9	4	0.9124	0.3668	1.6772	3.7248	0.2143	0.0754	0.02	0.384	3.61
44	4	0.9156	0.3621	1.7838	3.7248	0.2175	0.0761	0.02	0.3784	3.61
47.1	4	0.9175	0.3575	1.8981	3.7248	0.2175	0.0761	0.02	0.3732	3.61
49.7	4	0.9181	0.3536	2	3.7248	0.2208	0.0769	0.02	0.369	3.62

Calculation of wind forces, as per clause 10.2 of IS 875 (Part 3), 2015:

$$F_z = C_{f,z} A_z p_d G$$

where

F_z = design peak along-wind load on the building structure at any height z
$C_{f,z}$ = drag force coefficient of the building structure, as per figure 4 of IS 875 (Part 3), 2015
p_d = design hourly mean wind pressure corresponding to V_{zd} (N/m²)
V_{zd} = design hourly mean wind speed at height z (in m/s)
A_z = effective frontal area of the building structure at any height z, in m²
G = gust factor

Wind overall force coefficient in X direction

h =	49.7	h/b =	1.893189
b =	15.95	a/b =	1.454175
a =	26.45		
C_{fx} =	1.2	(as per clause 7.4.2 and figure 4)	

Wind overall force coefficient in Z direction

$h =$ 49.7

$b =$ 15.95 $h/a =$ 1.301899

$a =$ 26.45 $b/a =$ 0.687675

$C_{fz} =$ 1.25(as per clause 7.4.2 and figure 4)

As an input of STAAD Pro CE software, wind load intensities at different heights have to be provided, and the corresponding effective frontal area will be taken care of by the software itself.

Wind intensities (kN/m²) at different levels (z) have been calculated (Table 4.97) along the X and Z directions, as per the STAAD Pro CE axis system, with the input provided to the STAAD Pro CE software.

Across-Wind Response

As per clause 10.3 of IS 875 (Part 3), 2015.

The across-wind load distribution on the building can be calculated as follows.

$$F_{z,c} = (3M_c/h^2)\ (z/h)$$

where

$F_{z,c}$ = across-wind load per unit height at height z

$M_c = 0.5\ g_h p_h\ b\ h^2\ (1.06 - 0.06k)\ \sqrt{(\eta C_{fz}/\beta)}$

where

TABLE 4.97
Wind forces at different levels along heights

Height (m)	C_{fx}	A_z	P_d	G	F_{zx} (kN/m²)	C_{fz}	F_{zz} (kN/m²)
3.7	1.2	1	0.21	4.33	1.09	1.25	1.14
6.8	1.2	1	0.3	4.08	1.47	1.25	1.53
9.9	1.2	1	0.38	3.93	1.79	1.25	1.87
13	1.2	1	0.42	3.83	1.93	1.25	2.01
16.1	1.2	1	0.47	3.76	2.12	1.25	2.21
19.2	1.2	1	0.5	3.71	2.23	1.25	2.32
22.3	1.2	1	0.54	3.67	2.38	1.25	2.48
25.4	1.2	1	0.58	3.65	2.54	1.25	2.65
28.5	1.2	1	0.6	3.62	2.61	1.25	2.72
31.6	1.2	1	0.61	3.6	2.64	1.25	2.75
34.7	1.2	1	0.63	3.6	2.72	1.25	2.84
37.8	1.2	1	0.67	3.61	2.9	1.25	3.02
40.9	1.2	1	0.69	3.61	2.99	1.25	3.11
44	1.2	1	0.71	3.61	3.08	1.25	3.2
47.1	1.2	1	0.71	3.61	3.08	1.25	3.2
49.7	1.2	1	0.74	3.62	3.21	1.25	3.35

g_h = peak factor = $2\sqrt{(2\ln(3600\,f_c))}$
p_h = hourly mean wind pressure at height h, in Pa
b = breadth of the structure normal to the wind, in m
h = height of the structure, in m
k = mode shape power exponent for representation of the fundamental mode shape, as represented by $-\Psi(z) = (z/h)^k$
fc = first-mode natural frequency of the building / structure in across-wind direction, in Hz

Peak factor (g_h): it is already known that the natural frequency against the first mode = f_c = 0.347 Hz

$$g_h = 2\sqrt{(2\ln(3600\,f_c))}$$
$$= 7.55$$

Cross-Wind Force Spectrum Coefficient (C_{fs})

From figure 11 of IS 875 (Part 3), 2015, C_{fs} for rectangular buildings has been calculated.

The turbulence intensity is nearly 0.2 at 2/3 h = 2/3 × 49.7 m = 33.13 m. Allowing for turbulence intensity of 0.2 at 2/3 h, the corresponding curve is constructed (refer to figure 11 of IS 875 (Part 3), 2015).All these parameters are obtained for different levels (z) and furnished in Table 4.98.

TABLE 4.98
Cross-wind force spectrum

z (m)	V_{hd}	f_c	Wind direction in X b_z	Wind direction in Z b_x	Wind direction in X $V_{hd}/f_c b_z$	Wind direction in Z $V_{hd}/f_c b_x$	From figure 11 in IS 875 (Part 3), 2015 C_{fsz}	C_{fsx}
3.7	18.5	0.347	26.45	15.95	2.01566	3.342578	0.00075	0.002
6.8	22.5	0.347	26.45	15.95	2.45147	4.065298	0.0009	0.0025
9.9	25	0.347	26.45	15.95	2.72386	4.516997	0.00092	0.004
13	26.5	0.347	26.45	15.95	2.88729	4.788017	0.00092	0.0042
16.1	28	0.347	26.45	15.95	3.05072	5.059037	0.00092	0.0045
19.2	29	0.347	26.45	15.95	3.15968	5.239717	0.00092	0.0048
22.3	30	0.347	26.45	15.95	3.26863	5.420397	0.00095	0.005
25.4	31	0.347	26.45	15.95	3.37759	5.601077	0.00095	0.0055
28.5	31.5	0.347	26.45	15.95	3.43206	5.691417	0.00095	0.0055
31.6	32	0.347	26.45	15.95	3.48654	5.781757	0.00095	0.006
34.7	32.5	0.347	26.45	15.95	3.54102	5.872097	0.00095	0.006
37.8	33.5	0.347	26.45	15.95	3.64997	6.052777	0.00098	0.007
40.9	34	0.347	26.45	15.95	3.70445	6.143117	0.00098	0.007
44	34.5	0.347	26.45	15.95	3.75893	6.233456	0.001	0.0072
47.1	34.5	0.347	26.45	15.95	3.75893	6.233456	0.001	0.0072
49.7	35	0.347	26.45	15.95	3.8134	6.323796	0.001	0.0075

As per clause 10.3.1b of IS 875 (Part 3), 2015:

k = mode shape power exponent for slender-framed-structure
(moment-resisting frame) = 0.5

As per Table 36 of IS 875 (Part 3), 2015:

ß = Damping coefficient of the building/structure for an Reinforced Concrete
structure = 0.02

Typical calculation of M_c and F_{zc} for z = 3.7 m:

M_{cx} = (0.5)(7.55)(0.21)(15.95)(49.7)^2(1.06 – (0.06)(0.5)) $\sqrt{((22/7)(0.002)/0.02)}$
    = 18034.71 kNm

Similarly:

M_{cz} = (0.5)(7.55)(0.21)(26.45)(49.7)^2(1.06 – (0.06)(0.5)) $\sqrt{((22/7)(0.00075)/0.02)}$
    = 18314.28 kNm
F_{zcx} = (3(18034.71)/(49.7)^2)(3.7/49.7) / (15.95) = 0.1 kN/m^2
F_{zcz} = (3(18314.28)/(49.7)^2)(3.7/49.7) / (26.5) = 0.06 kN/m^2

Detailed calculations of Mc and Fzc at different levels along the height have been computed and furnished in Table 4.99.

4.5.2.5   Wind Load Analysis Using Software

Wind definition has to be provided along the height of the building using calculations made so far, and software have provided outputs.

4.5.3   SEISMIC LOAD ANALYSIS USING SOFTWARE

Seismic definition has been provided along the height of the building using calculations made so far, and software have provided outputs.

4.5.4   DIFFERENT CHECKS

Different Results Need to Be Studied from the Output

1. The floor diaphragm at different levels, with the position of the center of mass.
2. A soft story check.
3. The mode, frequency and period of the building.
4. The peak story shear at different levels.
5. Calculation of the eccentricity, as per clause 7.8.2 of IS 1893 (Part 1), 2016, for each response spectrum load case.
6. Calculation of the torsional moment and the peak additional torsion.

TABLE 4.99
Values of M_c and F_{zc} (forces considering per unit height and per unit width)

z (m)	g_h	P_h	Wind direction in X b_z	Wind direction in Z b_x	h	k	Wind direction in X Cf_{sz}	Wind direction in Z Cf_{sx}	β	Wind direction in X M_{cz} (kNm)	Wind direction in Z M_{cx} (kNm)	Wind direction in X F_{xcz} per unit height	Wind direction in Z F_{zcx} per unit height	Wind direction in X F_{zcz}/b_z (kN/m²)	Wind direction in Z F_{zcx}/b_x (kN/m²)
3.7	7.55	0.21	26.45	15.95	49.7	0.5	0.00075	0.002	0.02	18,314.28	18,034.71	1.66	1.63	0.06	0.1
6.8	7.55	0.30	26.45	15.95	49.7	0.5	0.0009	0.0025	0.02	28,660.41	28,804.89	4.76	4.79	0.18	0.3
9.9	7.55	0.38	26.45	15.95	49.7	0.5	0.00092	0.004	0.02	36,704.34	46,151.79	8.88	11.17	0.34	0.7
13	7.55	0.42	26.45	15.95	49.7	0.5	0.00092	0.0042	0.02	40,567.96	52,269.57	12.89	16.61	0.49	1.04
16.1	7.55	0.47	26.45	15.95	49.7	0.5	0.00092	0.0045	0.02	45,397.48	60,545.11	17.86	23.82	0.68	1.49
19.2	7.55	0.50	26.45	15.95	49.7	0.5	0.00092	0.0048	0.02	48,295.19	66,522.04	22.66	31.21	0.86	1.96
22.3	7.55	0.54	26.45	15.95	49.7	0.5	0.00095	0.005	0.02	53,002.40	73,325.28	28.88	39.96	1.09	2.51
25.4	7.55	0.58	26.45	15.95	49.7	0.5	0.00095	0.0055	0.02	56,928.50	82,600.81	35.34	51.27	1.34	3.21
28.5	7.55	0.60	26.45	15.95	49.7	0.5	0.00095	0.0055	0.02	58,891.55	85,449.11	41.02	59.51	1.55	3.73
31.6	7.55	0.61	26.45	15.95	49.7	0.5	0.00095	0.006	0.02	59,873.08	90,736.17	46.23	70.07	1.75	4.39
34.7	7.55	0.63	26.45	15.95	49.7	0.5	0.00095	0.006	0.02	61,836.13	93,711.12	52.44	79.46	1.98	4.98
37.8	7.55	0.67	26.45	15.95	49.7	0.5	0.00098	0.007	0.02	66,792.52	107,646.22	61.70	99.44	2.33	6.23
40.9	7.55	0.69	26.45	15.95	49.7	0.5	0.00098	0.007	0.02	68,786.32	110,859.54	68.75	110.80	2.6	6.95
44	7.55	0.71	26.45	15.95	49.7	0.5	0.001	0.0072	0.02	71,498.73	115,691.00	76.88	124.40	2.91	7.8
47.1	7.55	0.71	26.45	15.95	49.7	0.5	0.001	0.0072	0.02	71,498.73	115,691.00	82.29	133.16	3.11	8.35
49.7	7.55	0.74	26.45	15.95	49.7	0.5	0.001	0.0075	0.02	74,519.80	123,065.78	90.51	149.47	3.42	9.37

7. Calculation of the base actions (Fx, Fy, Fz, Mx, My and Mz) for each mode.
8. The mass participation factor (in percent);the number of modes to be used, as per clause 7.7.5.2 of IS 1893 (Part 1), 2016; the mass participation factor should be more than 90 percent.
9. The base shear factor (as per clause 7.7.3a of IS 1893 (Part 1), 2016) also has to be calculated, as per the equivalent static method, and if it is higher than the response spectrum base shear then the response spectrum base shear is to be multiplied by the factor V_b/V_B.
10. Different irregularity checks to be done, as per IS 1893 (Part 1), 2016:

   a. the torsion irregularity check;
   b. the geometry irregularity check;
   c. the mass irregularity check;
   d. irregular modes of oscillation.

4.5.5 DESIGN OF BEAMS, COLUMNS AND PILE CAPS USING SOFTWARE

Computer analysis of the building structure under all the relevant loads, such as the dead, live, wind and seismic loads, have been carried out. The design load combinations as per the limit state theory of IS 456, 2000, have been given as inputs. Then the command to make designs for the beams, columns, etc. as per IS 456, 2000, and IS 13920, 2016, has been given. All the beams and columns have been designed by software (STAAD Pro CE), and then detailed reinforcement drawings are prepared using AUTOCAD. The foundation design (for the pile caps) has also been designed using the software (STAAD Foundation Advanced CE).

During the preparation of the reinforcement drawings, IS 2505 (the code of practice for the bending and fixing of bars and wires for concrete reinforcement) and SP 34 (the handbook on concrete reinforcement detailing) have been followed, applying good professional practice. Ductile detailing as per IS 13920, 2016, has been carriedout, and the recommendations regarding the proper confining reinforcement, and other stipulations to make better resistance under seismic activities, are strictly followed.

4.5.6 WORKING DRAWINGS OF SLABS, BEAMS, COLUMNS AND FOUNDATIONS

Working drawings of the building showing the reinforcement detailing etc. are available in Figures 4.48, 4.49, 4.50, 4.51, 4.52, 4.53, 4.54 and 4.55.

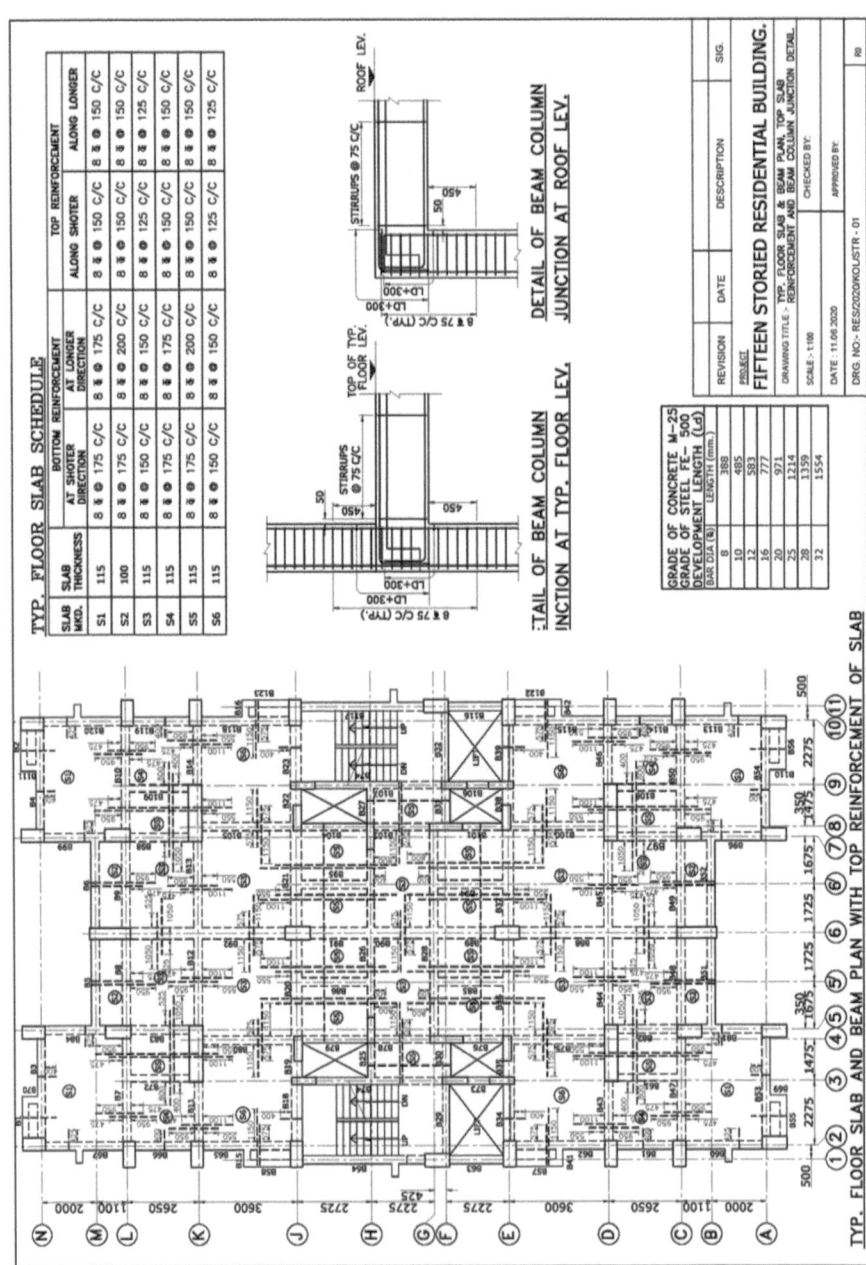

FIGURE 4.48 Reinforcement details of slab (top reinforcement only) for 15-storied building

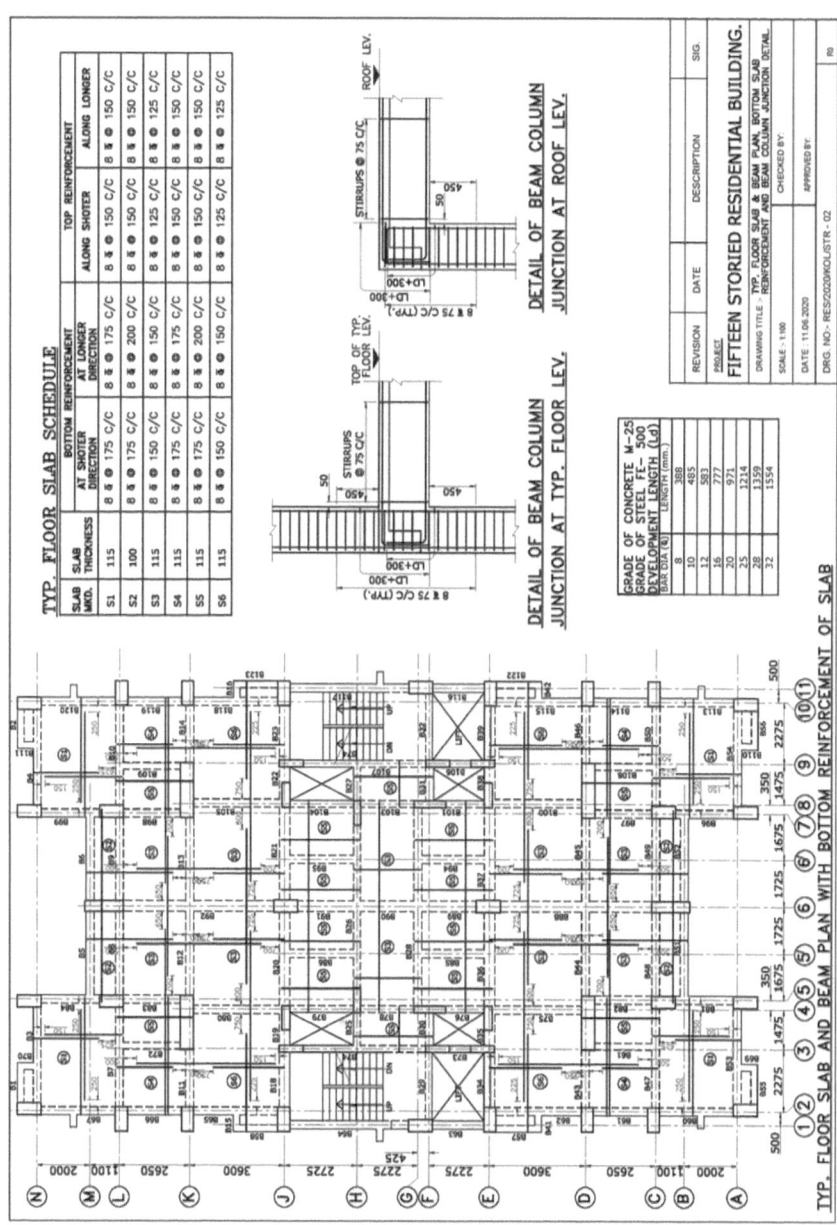

FIGURE 4.49 Reinforcement details of slab (bottom reinforcement only) for 15-storied building

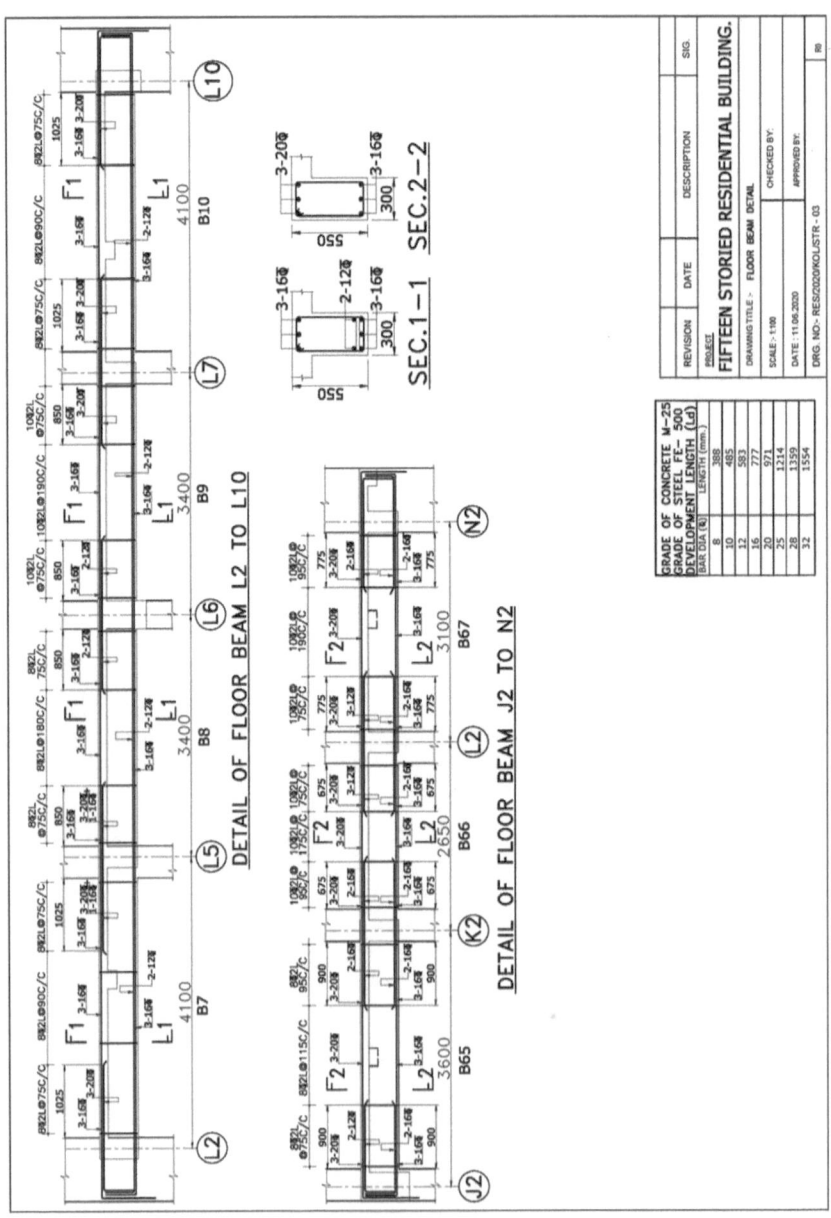

FIGURE 4.50 Reinforcement details of beams for 15-storied building

TYP. FLOOR BEAM SCHEDULE

BEAM NOS.	SIZE B	D	BOTTOM REINFORCEMENT LEFT	MID SPAN	RIGHT	TOP REINFORCEMENT LEFT	MID SPAN	RIGHT	SHEAR STIRRUPS LEFT	MID SPAN	RIGHT
B1	250	250	3-12T	3-12T	3-12T	3-12T	3-12T	3-12T	8T2L@185 C/C	8T2L@185 C/C	8T2L@185 C/C
B2	250	250	3-12T	3-12T	3-12T	3-12T	3-12T	3-12T	8T2L@185 C/C	8T2L@185 C/C	8T2L@185 C/C
B3	300	550	+3-16T 2-12T	+3-16T 2-12T	3-12T	+3-16T 2-20T+1-16T	3-16T	+3-16T 2-20T+1-16T	8T2L@75 C/C	8T2L@115 C/C	8T2L@75 C/C
B4	300	550	+3-16T 2-12T	+3-16T 2-12T	+3-16T 2-12T	+3-16T 2-20T+1-16T	3-16T	3-T25	8T2L@75 C/C	8T2L@115 C/C	8T2L@75 C/C
B5	300	550	3-16T	3-16T	3-16T	+3-16T 2-20T+1-16T	3-16T	+3-16T +2-12T	8T2L@190 C/C	8T2L@190 C/C	8T2L@190 C/C
B6	300	550	3-16T	3-16T	3-16T	+3-16T +2-12T	3-16T	3-16T	8T2L@190 C/C	8T2L@300 C/C	8T2L@300 C/C
B7	300	550	+3-16T 2-12T	+3-16T 2-12T	+3-16T 2-12T	+3-16T +3-20T	3-16T	+3-16T 2-20T+1-16T	8T2L@75 C/C	8T2L@90 C/C	8T2L@75 C/C
B8	300	550	+3-16T 2-12T	+3-16T 2-12T	+3-16T 2-12T	+3-16T 2-20T+1-16T	3-16T	+3-16T +2-12T	8T2L@75 C/C	8T2L@180 C/C	8T2L@75 C/C
B9	300	550	+3-16T 2-12T	+3-16T 2-12T	+3-16T 2-12T	+3-16T +3-20T	3-16T	+3-16T +3-20T	10T2L@75 C/C	10T2L@190 C/C	10T2L@75 C/C
B10	300	550	+3-16T 2-12T	+3-16T 2-12T	+3-16T 2-12T	+3-16T +3-20T	3-16T	+3-16T +3-20T	8T2L@75 C/C	8T2L@90 C/C	8T2L@75 C/C
B11	300	550	+3-20T 2-12T	+3-20T 2-12T	+3-20T 2-12T	+3-20T 2-16T+1-20T	3-20T	+3-20T +3-20T	10T2L@75 C/C	10T2L@110 C/C	10T2L@75 C/C
B12	300	600	+3-20T 2-12T	+3-20T 2-12T	+3-20T 2-12T	+3-20T 2-16T+1-20T	3-20T	+3-20T	8T2L@125 C/C	8T2L@130 C/C	8T2L@145 C/C
B13	300	600	+3-20T 2-12T	+3-20T 2-12T	+3-12T	3-20T	3-20T	+3-20T +3-20T	8T2L@205 C/C	8T2L@195 C/C	8T2L@175 C/C
B14	300	550	+3-20T 2-12T	+3-20T 2-12T	3-12T	+3-20T +3-20T	3-20T	+3-20T +3-20T	10T2L@75 C/C	10T2L@135 C/C	10T2L@75 C/C
B15	250	250	3-12T	3-12T	3-12T	3-12T	3-12T	3-12T	8T2L@105 C/C	8T2L@105 C/C	8T2L@105 C/C
B16	250	250	3-12T	3-12T	3-12T	3-12T	3-12T	3-12T	8T2L@105 C/C	8T2L@105 C/C	8T2L@105 C/C
B17	250	250	3-12T	3-12T	3-12T	3-12T	3-12T	3-12T	8T2L@105 C/C	8T2L@105 C/C	8T2L@105 C/C
B18	400	600	+3-20T 2-12T	3-20T	3-20T	+3-20T 2-16T+1-12T	3-20T	+3-20T 2-16T+1-12T	10T2L@75 C/C	10T2L@75 C/C	10T2L@75 C/C
B19	400	600	+3-20T 2-12T	+3-20T 2-12T	3-12T	+3-20T 2-16T+1-12T	+2-16T+1-12T	+3-20T 2-16T+1-12T	8T2L@160 C/C	8T2L@130 C/C	8T2L@120 C/C
B20	400	600	3-20T	+3-20T 2-12T	3-20T	+3-20T 2-16T+1-12T	3-20T	+2-16T+1-20T	8T2L@85 C/C	8T2L@105 C/C	8T2L@75 C/C

GRADE OF CONCRETE M-25
GRADE OF STEEL FE- 500
DEVELOPMENT LENGTH (Ld)

BAR DIA(%)	LENGTH (mm)
8	388
10	485
12	583
16	777
20	971
25	1214
28	1359
32	1554

REVISION	DATE	DESCRIPTION	SIG

PROJECT
FIFTEEN STORIED RESIDENTIAL BUILDING.
DRAWING TITLE :- TYP. FLOOR BEAM SCHEDULE
SCALE :- | CHECKED BY
DATE 11.06.2020 | APPROVED BY
DRG. NO:- RES/2020/KOL/STR - 04

FIGURE 4.51 Reinforcement schedule of beams for 15-storied building

Note: Reinforcement chart of remaining beams may be represented in a similar manner.

TYP. FLOOR BEAM SCHEDULE

BEAM NOS.	SIZE B	SIZE D	BOTTOM REINF. LEFT	BOTTOM REINF. MID SPAN	BOTTOM REINF. RIGHT	TOP REINF. LEFT	TOP REINF. MID SPAN	TOP REINF. RIGHT	SHEAR STIRRUPS LEFT	SHEAR STIRRUPS MID SPAN	SHEAR STIRRUPS RIGHT
B21	400	600	3-20T	3-20T	3-20T / +3-12T	+3-20T / 2-16T+1-20T	3-20T	+3-20T / 2-16T+1-12T	10T2L@80 C/C	10T2L@95 C/C	10T2L@75 C/C
B22	400	600	+3-20T / 3-12T	+3-20T / 3-12T	+3-20T / 3-12T	+3-20T / 2-16T+1-12T	3-20T	+3-20T / 2-16T+1-12T	8T2L@90 C/C	8T2L@100 C/C	8T2L@105 C/C
B23	400	600	+3-20T / 3-12T	3-20T	3-20T	+3-20T / 2-16T+1-12T	3-20T	+3-20T / 2-16T+1-12T	10T2L@75 C/C	10T2L@75 C/C	10T2L@75 C/C
B24	250	250	3-12T	3-12T	3-12T	3-12T	3-12T	3-12T	8T2L@95 C/C	8T2L@95 C/C	8T2L@95 C/C
B25	300	550	+3-20T / 2-12T	+3-20T / 2-12T	+3-20T / 2-12T	+3-20T / 2-20T+1-16T	+3-20T / 2-20T+1-16T	+3-20T / 2-20T+1-16T	8T2L@180 C/C	8T2L@175 C/C	8T2L@175 C/C
B26	300	550	+3-20T / 2-12T	+3-20T / 2-12T	+3-20T / 2-12T	+3-20T / 2-20T+1-16T	+3-20T / 2-20T+1-16T	+3-20T / 2-20T+1-16T	8T2L@75 C/C	8T2L@115 C/C	8T2L@75 C/C
B27	300	550	+3-20T / 2-12T	+3-20T / 2-12T	+3-20T / 2-12T	+3-20T / 2-20T+1-16T	+3-20T / 2-20T+1-16T	+3-20T / 2-20T+1-16T	8T2L@170 C/C	8T2L@175 C/C	8T2L@180 C/C
B28	300	550	3-20T	3-20T	3-20T	+3-20T / 2-12T	3-20T	+3-20T / 2-12T	10T2L@75 C/C	10T2L@110 C/C	10T2L@75 C/C
B29	300	550	3-20T	3-20T	3-20T	+3-20T / 2-16T+1-12T	+3-20T / 2-16T+1-12T	+3-20T / 2-16T+1-12T	10T2L@85 C/C	10T2L@190 C/C	10T2L@85 C/C
B30	300	550	+3-16T / 2-12T	+3-16T / 2-12T	+3-16T / 2-12T	+3-20T / 2-16T+1-12T	3-20T	3-20T	10T2L@75 C/C	10T2L@75 C/C	10T2L@75 C/C
B31	300	550	3-20T	3-20T	3-20T	3-20T	3-20T	3-20T	10T2L@75 C/C	10T2L@75 C/C	10T2L@75 C/C
B32	300	550	3-20T	3-20T	3-20T	3-20T	3-20T	+3-20T / 2-16T+1-12T	10T2L@100 C/C	10T2L@190 C/C	10T2L@85 C/C
B33	250	250	3-12T	3-12T	3-12T	3-12T	3-12T	3-12T	8T2L@90 C/C	8T2L@90 C/C	8T2L@90 C/C
B34	400	600	3-20T	3-20T	+3-20T / 2-16T+1-12T	+3-20T / 2-16T	3-20T	+3-20T / 2-20T	10T2L@85 C/C	10T2L@85 C/C	10T2L@95 C/C
B35	400	600	+3-20T / 2-T16+1-T12	+3-20T / 2-16T+1-12T	+3-20T / 2-16T+1-12T	+3-20T / 2-20T	+3-20T / 2-20T	+3-20T / 2-20T	8T2L@90 C/C	8T2L@100 C/C	8T2L@105 C/C
B36	400	600	+3-20T / 2-T16+1-T12	3-20T	3-20T	+3-20T / 2-20T	+3-20T / 2-20T	+3-20T / 2-20T	10T2L@95 C/C	10T2L@100 C/C	10T2L@90 C/C
B37	400	600	3-20T	3-20T	3-20T	+3-20T / 2-20T	3-12T	+3-20T / 2-20T	10T2L@100 C/C	10T2L@140 C/C	10T2L@100 C/C
B38	400	600	+3-20T / 3-12T	+3-20T / +3-12T	+3-20T / 3-12T	+3-20T / 2-20T	+3-20T / 2-20T	+3-20T / 2-20T	8T2L@195 C/C	8T2L@170 C/C	8T2L@145 C/C
B39	400	600	+3-20T / 3-12T	3-20T	+3-20T / 3-12T	+3-20T / 2-20T	3-20T	+3-20T / 2-20T	10T2L@75 C/C	10T2L@75 C/C	10T2L@75 C/C

GRADE OF CONCRETE M-25
GRADE OF STEEL FE- 500
DEVELOPMENT LENGTH (Ld)

BAR DIA (Φ)	LENGTH (mm)
8	388
10	485
12	583
16	777
20	971
25	1214
28	1359
32	1554

REVISION	DATE	DESCRIPTION	SIG.

PROJECT
FIFTEEN STORIED RESIDENTIAL BUILDING.
DRAWING TITLE :- TYP. FLOOR BEAM SCHEDULE
SCALE - 1:30 | CHECKED BY:
DATE 11.06.2020 | APPROVED BY:
DRG. NO:- RES/2020/KOL/STR - 05 | R0

FIGURE 4.52 Reinforcement schedule of beams for 15-storied building

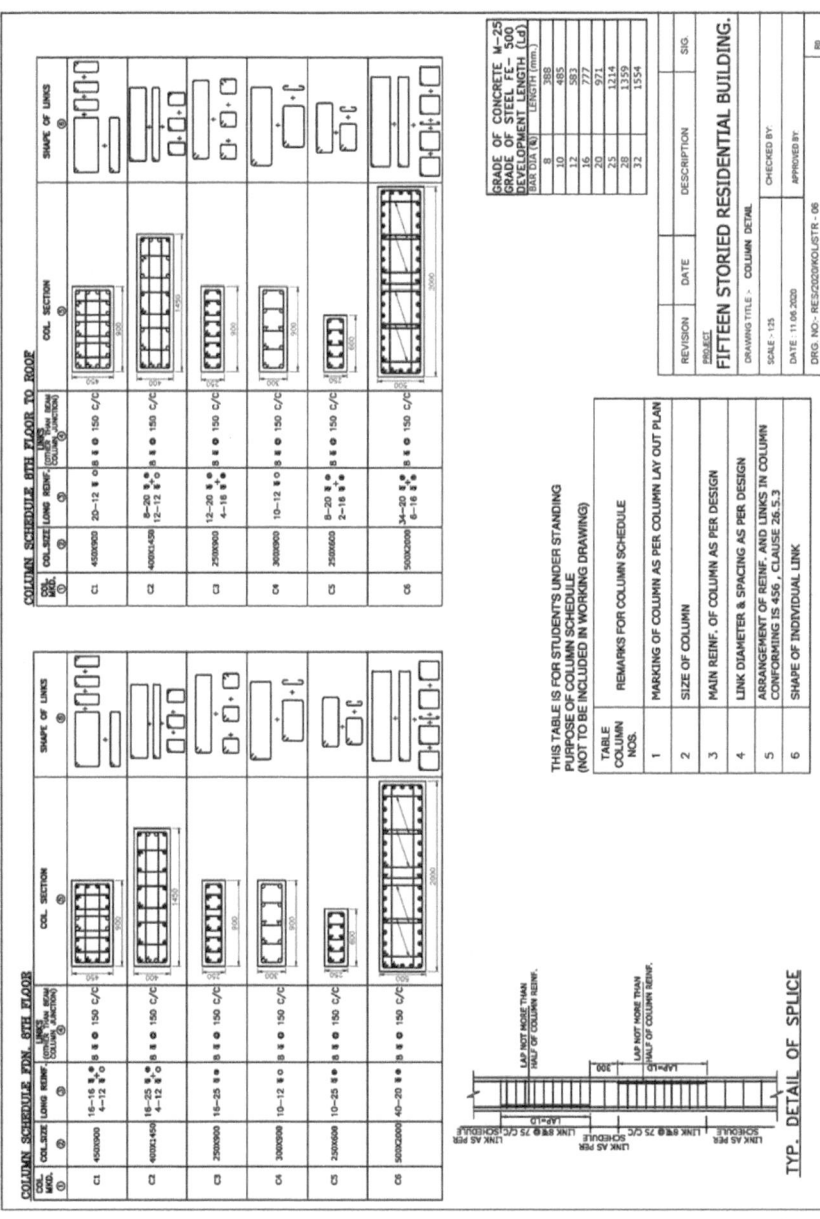

FIGURE 4.53 Reinforcement details of columns for 15-storied building

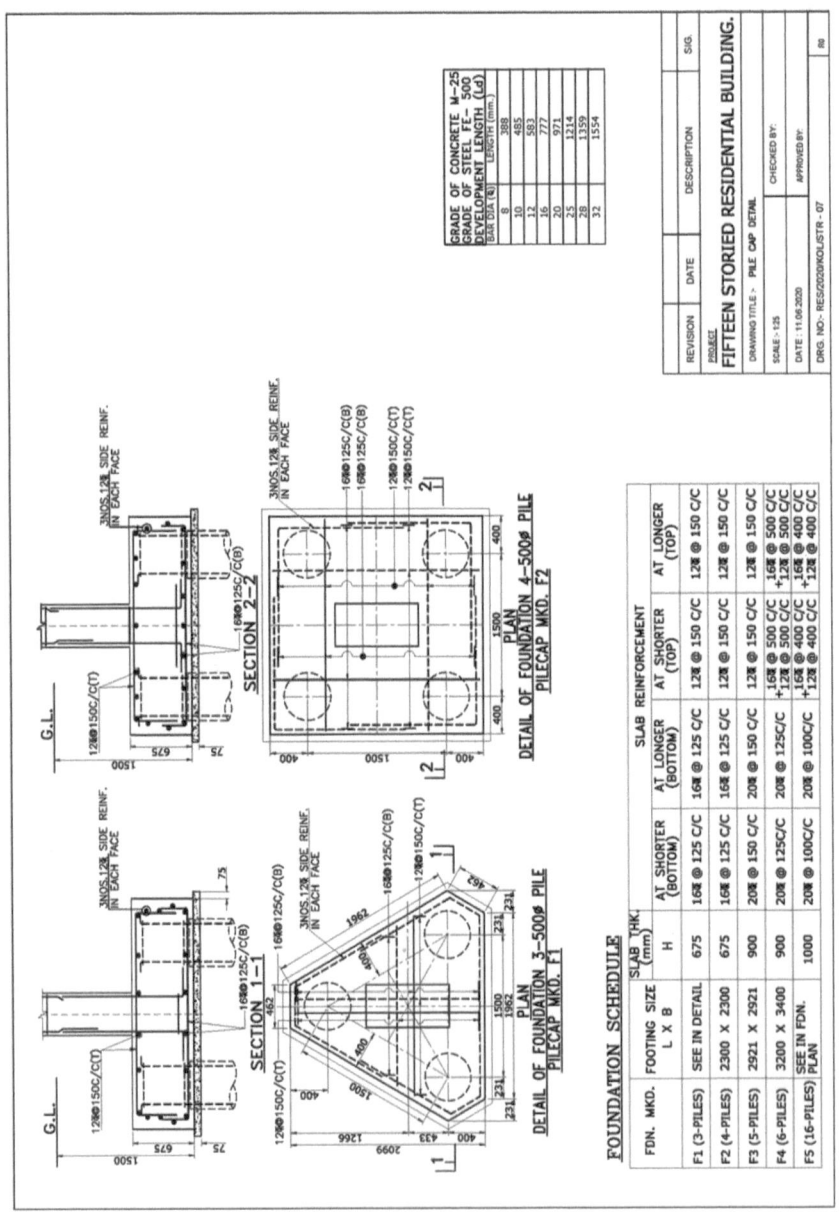

FIGURE 4.54   Reinforcement details of pile caps for 15-storied building

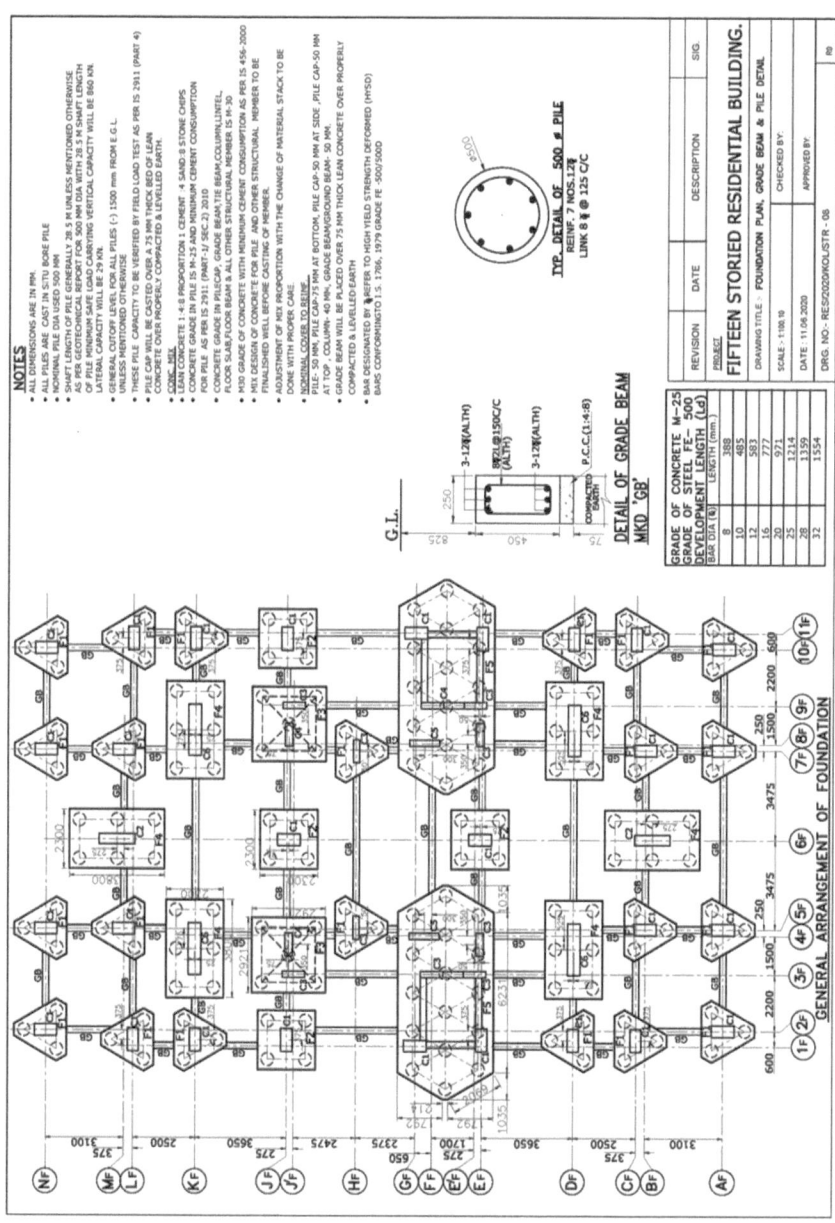

FIGURE 4.55 Layout of foundation (pile caps) including grade beams and piles and detail of reinforcements for 15-storied building

5 Comparison of Basic parameters stipulated for wind and seismic analysis, as per IS, IBC, ASCE, ACI, EN and BS Codes

5.1 PREAMBLE

Codes of practices, in different formats, have been developed in different countries, but, basically, they are all similar in concept. Therefore, an exercise is undertaken here to assess the design loads, specially wind and seismic loads. Typical steps in calculating the different parameters are shown, and a comparison is made of the basic parameters stipulated in Indian, American and British codes. *It is essential for the reader to have all the relevant codes of practices and handbooks for Reinforced Concrete design etc. as ready references while going through this chapter.*

Basic Data

A six-storied Reinforced Concrete-framed residential building is shown as per the plan and sectional elevation in Figures 5.1 and 5.2.

- Building to be constructed at Reston, Virginia, United States, ZIP code 20190
- Grade of concrete used = M25 and grade of steel used = Fe500 TMT deformed bars
- Typical floor height = 3,100 mm
- Plinth height = 600 mm
- Foundation depth = 1,500 mm

The design and analysis of the building is based on IBC 2012, ASCE 7–10 and ACI 318 codes.

The building will be constructed at ZIP code 20190 – Reston, Virginia.

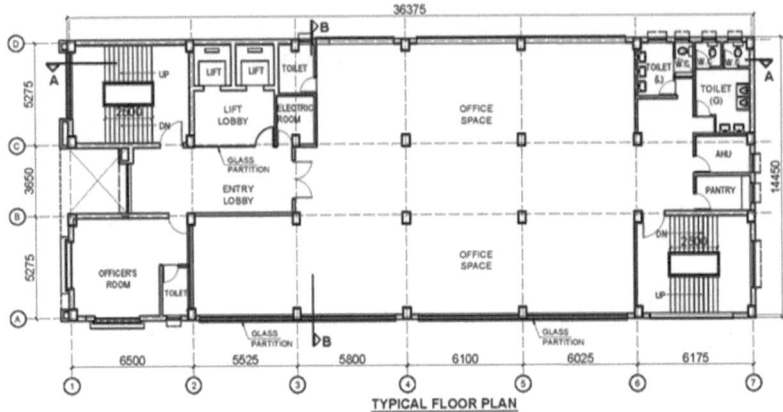

FIGURE 5.1   Plan of a six-storied building

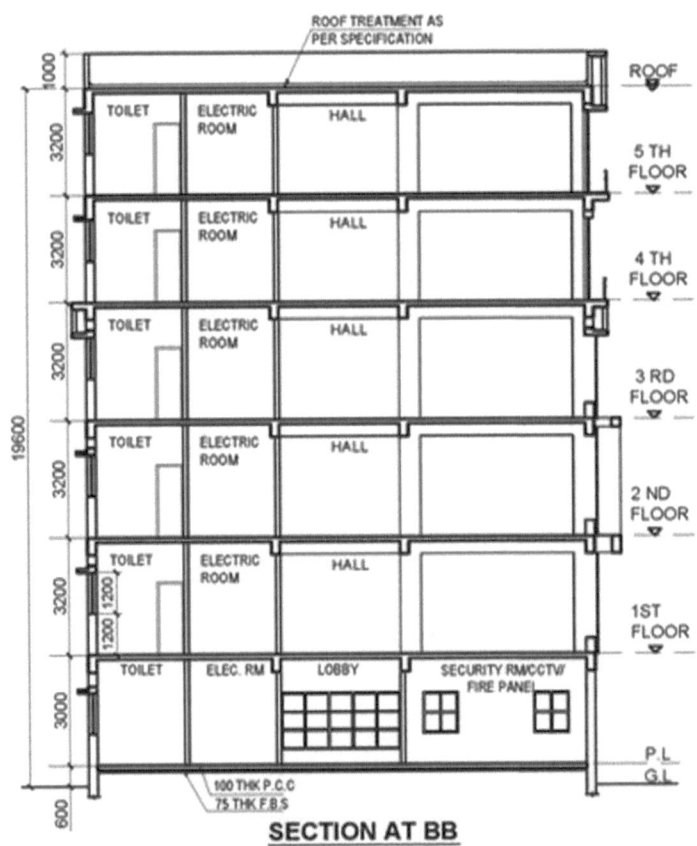

FIGURE 5.2   Sectional elevation of a six-storied building

DEAD LOAD AND LIVE LOAD

The minimum loads considered in Chapter 4 as per Indian code are slightly greater than the minimum load mentioned in ASCE 7–10, so therefore they satisfy the loading criteria as per ASCE 7–10. Hence, the dead loads and live loads already calculated and considered in Chapter 4 can be used in this exercise without any change.

5.2 WIND LOAD ANALYSIS

The wind load is calculated on the basis of the data furnished in Table 5.1.

TABLE 5.1
Basic data considered for wind load analysis

Location	ZIP code 20190, latitude 38.9597, longitude 77.3374
Occupancy	Miscellaneous office building
Dimensions	36.375 m (b) × 14.45 m (d) in plan; roof height 23.6 m (h)
Terrain	Flat

The design wind pressure for enclosed and partially enclosed buildings as per clause 27.4.1 of ASCE 7–10 is:

$$p = q \, G \, C \, p - q_i \, (G \, Cp_i) \qquad (1)$$

where

G = gust effect factor
C_p = external pressure coefficient
C_{pi} = internal pressure coefficient
q = velocity pressure, in psf

The design wind pressure for enclosed and partially enclosed buildings as per clause 29.3.2 of ASCE 7–10 is:

$$q = 0.613 K_z K_{zt} K_d V^2 \qquad (2)$$

where

q is in (N/m2) and V is in m/s
$q = q_h$ for leeward walls, side walls and roofs, evaluated at roof mean height, h
$q = q_z$ for windward walls, evaluated at height z
$q_i = q_h$ for negative internal pressure, $(-G \, C_{pi})$ evaluation, and q_z for positive internal pressure evaluation $(+G \, C_{pi})$ of partially enclosed buildings, but can be taken as q_h for a conservative value
K_z = velocity pressure coefficient
K_{zt} = topographic factor

K_d = wind directionality factor
V = basic wind speed

Risk Category
From table 1.5-1 of ASCE 7–10, the risk category of the building is under category II. As per clause 26.2 of ASCE 7–10, the building may be considered partially enclosed.

Basic Wind Speed
As per figures 26.5-1A, B and C of ASCE 7–10:

= 115 mph = 51.4 m/s

Exposure category
As per clause 26.7.3 of ASCE 7–10, the exposure category is under category B.

Wind directionality factor (K_d)
As per table 26.6-1 of ASCE 7–10, the main wind-force-resisting system of the building = 0.85.

Topographic factor (K_{zt})
As the building site is flat, K_{zt} = 1.0.

Gust Effect Factor (G)
As per clause 26.9.1, it is considered as a rigid building: G = 0.85.

Internal Pressure Coefficient (GC_{pi}) = product of the internal pressure coefficient and the gust effect factor, to be used indetermination of wind loads for buildings.
As per table 26.11-1, the enclosure classification is for partially enclosed buildings:

$$GC_{pi} = +0.55/–0.55$$

Velocity Pressure Coefficient (K_z)
As per table 27.3-1 of ASCE 7–10, K_z varies with height, and the data are furnished in Table 5.2.

Velocity Pressure (reference equation 2)
Referring equation 2

$$q = 0.613 \ K_z \ K_{zt} \ K_d \ V^2$$

The Velocity-pressure at different heights is computed in Table 5.3, using the data furnished in Table 5.2.

TABLE 5.2
Velocity pressure coefficient
(K_z) along heights

Height (m)	K_z
3.6	0.57
6.8	0.64
10	0.72
13.2	0.78
16.4	0.83
19.6	0.87
23.6	0.91

TABLE 5.3
Velocity pressure (q) along heights

Height (m)	K_z	K_d	K_{zt}	V	q (kN/m²)
3.6	0.57	0.85	1	51.4	0.785
6.8	0.64	0.85	1	51.4	0.881
10	0.72	0.85	1	51.4	0.991
13.2	0.78	0.85	1	51.4	1.074
16.4	0.83	0.85	1	51.4	1.143
19.6	0.87	0.85	1	51.4	1.198
23.6	0.91	0.85	1	51.4	1.253

External Pressure Coefficient (C_p) for Walls

The length/width ratio of the building (L/B) = 36.375/14.45 = 2.5.

As per figure 27.4-1 of ASCE 7–10 and clause 27.4.1, the external pressure coefficient for walls is calculated and furnished in Table 5.4.

Design Wind Pressure for Main Wind Frame Resisting System

$$p = q \ G \ C_p - q_i \ (GC_{pi})$$

where, p is in N/m²

The design wind pressure on walls is computed and furnished in Tables 5.5, 5.6 and 5.7, using the data furnished in Table 5.4.

TABLE 5.4
Calculation of external pressure coefficient (C_p) for wall

Wall pressure coefficients, Cp

Surface	L/B	C_p
Windward wall	All values	0.8
Leeward wall	2.5	−0.275
Side walls	All values	−0.7

TABLE 5.5
Design wind pressure on the windward wall

Height (m)	q (kN/m²)	G	(+GC$_{pi}$)	(−GC$_{pi}$)	C_p	(+GC$_{pi}$)	(−GC$_{pi}$)
3.6	0.785	0.85	0.55	−0.55	0.8	−0.155	1.223
6.8	0.881	0.85	0.55	−0.55	0.8	−0.09	1.288
10	0.991	0.85	0.55	−0.55	0.8	−0.015	1.363
13.2	1.074	0.85	0.55	−0.55	0.8	0.041	1.419
16.4	1.143	0.85	0.55	−0.55	0.8	0.088	1.466
19.6	1.198	0.85	0.55	−0.55	0.8	0.125	1.504
23.6	1.253	0.85	0.55	−0.55	0.8	0.163	1.541

TABLE 5.6
Wind pressure on the leeward wall

Height (m)	q (kN/m²)	G	(+GC$_{pi}$)	(−GC$_{pi}$)	C_p	(+GC$_{pi}$)	(−GC$_{pi}$)
3.6	0.785	0.85	0.55	−0.55	−0.275	−0.95	0.4
6.8	0.881	0.85	0.55	−0.55	−0.275	−0.95	0.4
10	0.991	0.85	0.55	−0.55	−0.275	−0.95	0.4
13.2	1.074	0.85	0.55	−0.55	−0.275	−0.95	0.4
16.4	1.143	0.85	0.55	−0.55	−0.275	−0.95	0.4
19.6	1.198	0.85	0.55	−0.55	−0.275	−0.95	0.4
23.6	1.253	0.85	0.55	−0.55	−0.275	−0.95	0.4

5.3   SEISMIC LOAD ANALYSIS

As per IBC 2012 and ASCE 7–10, the seismic parameters are calculated as follows.

Input to STAAD Pro CE software as seismic definition
Code IBC 2012, ASCE 7–10
Risk category, from Table 1.5-1, is II
Importance factor for seismic load = 1.0, as per table 1.5-2

TABLE 5.7
Design wind pressure on the side walls

Height (m)	q (kN/m2)	G	$(+GC_{pi})$	$(-GC_{pi})$	C_p	$(+GC_{pi})$	$(-GC_{pi})$
3.6	0.785	0.85	0.55	−0.55	−0.7	−1.435	−0.056
6.8	0.881	0.85	0.55	−0.55	−0.7	−1.435	−0.056
10	0.991	0.85	0.55	−0.55	−0.7	−1.435	−0.056
13.2	1.074	0.85	0.55	−0.55	−0.7	−1.435	−0.056
16.4	1.143	0.85	0.55	−0.55	−0.7	−1.435	−0.056
19.6	1.198	0.85	0.55	−0.55	−0.7	−1.435	−0.056
23.6	1.253	0.85	0.55	−0.55	−0.7	−1.435	−0.056

ZIP code − 20190

The corresponding S_S and S_i values, as per figures 22-1 to 22-6, are calculated by the software

The response modification factor (R), as per table 12.2-1 considering "special re-inforced concrete systems", R = 5

The occupancy importance factor (I) (as per clause 1604.5 of IBC 2012 and table 11.5-1 of ASCE 7–10), I = 1

The long-period transition period (T_L), in seconds, (as per clause 11.4.5 and chapter 22, figures 22-12 to 22-16, of ASCE 7–10): T_L = 12 (calculated by the software)

For the site class (SCL), enter 1 through 6 in place of A through F (clause 1613.3.2 of IBC 2012, section 20.3 of ASCE 7–10) = 4 (i.e., D) (as per 11.4.2)

C_t = time period (as per table 12.8-2 of ASCE 7–10) = 0.0466 for concrete moment-resisting frames

Exponent value (x) used in equation 12.8-7 in clause 12.8.2.1 of ASCE 7–10 (as per ASCE 7–10 table 12.8-2) = 0.9

Short-period site coefficient at 0.2s (as per clause 1613.3.3 of IBC 2012 and section 11.4.3 of ASCE 7–10): F_a = 1.6

Long-period site coefficient at 1.0s (as per clause 1613.3.3 of IBC 2012 and section 11.4.3 of ASCE 7–10): F_v = 2.4

Using the above-mentioned data, a seismic analysis is carried out using STAAD Pro CE software.

5.4 NUMERICAL EXAMPLE OF WIND AND SEISMIC LOAD ANALYSIS

The design and analysis of the building is carried out as per the EN 1991-1-1: 2002, EN 1991-1-4: 2005, EN 1998-1: 2004 and BS 8110 codes.

Basic Data

- Building site at Inden, Nordrhein Westfalen, Germany, ZIP code 52459
- Grade of concrete used = M25 and grade of reinforcing steel used = Fe500 TMT deformed bars
- Typical floor-to-floor height = 3,100 mm
- Plinth height = 600 mm
- Foundation depth = 1,500 mm
- Building plan and sectional elevation as per Figures 5.3 and 5.4

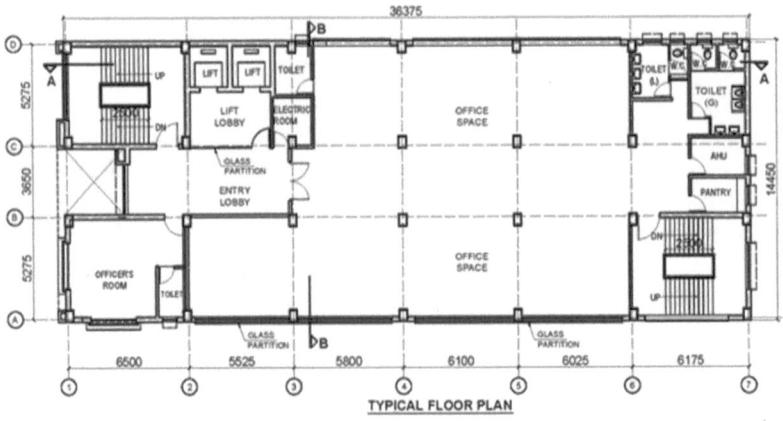

FIGURE 5.3   Plan of a six-storied building

The minimum dead and live loads considered in Chapter 4 as per the Indian code are slightly greater than the minimum load mentioned in EN 1991-1-1: 2002. They therefore satisfy the loading criteria of EN 1991-1-1: 2002. Hence, the dead and live loads already taken into account in Chapter 4 are used again in this exercise without change.

However, the wind load is calculated as per EN 1991-1-4: 2005, on the basis of the basic building data as furnished in Table 5.8.

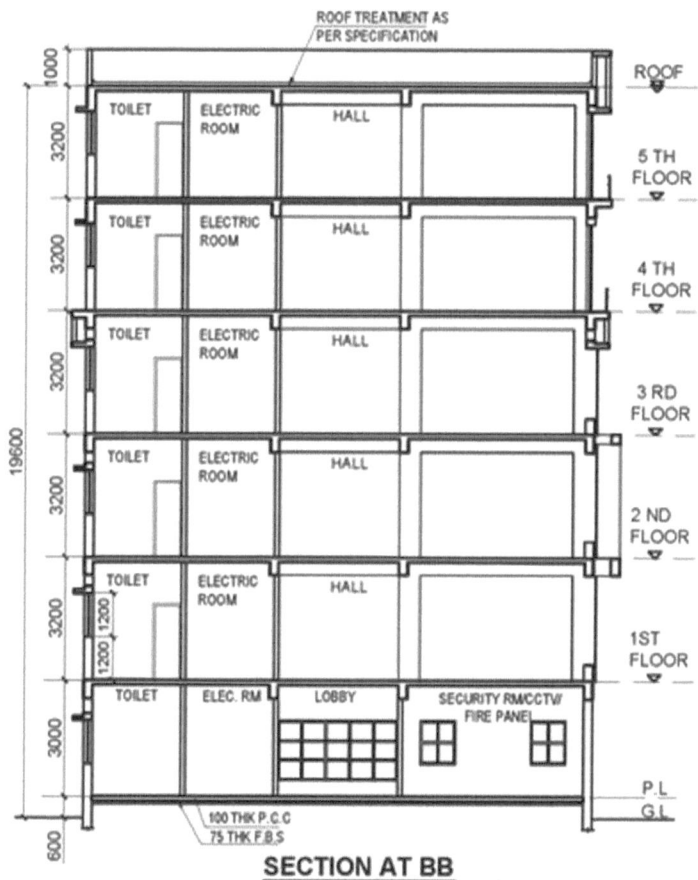

FIGURE 5.4   Sectional elevation of a six-storied building

TABLE 5.8
Basic data for wind load assessment

Location	Inden, ZIP code 52459, Germany
Occupancy	Miscellaneous office building
Dimensions	36.375 m (b) × 14.45 m(d) in plan; roof height 23.6 m (h)
Terrain	Flat

Basic Wind Velocity

As per clause 4.2 of EN 1991-1-4: 2010:

$$V_b = C_{dir} C_{season} V_{b,0} \qquad (A)$$

where

V_b = basic wind velocity, in m/s
C_{dir} = directional factor
C_{season} = seasonal factor
$V_{b,0}$ = fundamental value of the basic wind velocity (DIN National Annex for EN 1991-1-4: 2010)

As per clause 4.2 of EN 1991-1-4: 2010, and notes 2 and 3, C_{dir} = 1.0, C_{season} = 1.0 and $V_{b,0}$ = 25 m/s (from the National Annex). Therefore:

$$V_b = 25 \text{ m/s}$$

As per clause 4.3.1 of EN 1991-1-4: 2010:

$$v_m(z) = c_r(z)\, c_o(z)\, V_b \qquad \textbf{(B)}$$

where

$v_m(z)$ = mean wind velocity, m/s
$c_r(z)$ = roughness factor
$c_o(z)$ = orography factor

As per clause 4.3.2 of EN 1991-1-4: 2010, terrain factors are calculated:

$$c_r(z) = K_r \ln(z/z_0) \text{ for } z_{min} \leq z \leq z_{max} \text{ ---- (C)}$$

$$c_r(z) = c_r(z_{min}) \text{ for } z \leq z_{min} \text{ ----------(D)}$$

where

z_0 = roughness length, in meters
K_r = terrain factor, depending on the roughness length, z_0

$$K_r = 0.19(z_0/z_{0,II})^{0.07} \qquad \textbf{(E)}$$

where

$z_{0,II}$ = 0.05 (terrain category II, as per Table 4.1)
z_{min} = minimum height = 2 m (terrain category II, as per Table 4.1)
z_{max} = maximum height, taken as 200 m

Terrain orography may be considered as per clause 4.3.3 of EN 1991-1-4: 2010. The orography effect is neglected as per note 2 of clause 4.3.3 of EN 1991-1-4: 2010.

$$co(z) = 1.0$$

Wind Turbulence

Turbulence intensity is calculated as per clause 4.3.4 of EN 1991-1-4: 2010:

$$Iv(z) = K_l/(co(z).ln(z/z_0))$$

where

K_l is the turbulence factor = 1.0

Basic Velocity and Peak Velocity Pressure

Basic velocity pressure is calculated as per clause 4.5 of EN 1991-1-4: 2010:

$$q_b = 0.5\rho_{air}V_b^2 \qquad\qquad\text{(F)}$$

where

q_b = design wind pressure, in Pa
ρ_{air} = density of air (1.25 kg/m³)
V_b = basic wind velocity, in m/s

$$q_{b\,=}\ 0.39\ kPa$$

Peak velocity pressure

$$q_p(z) = 0.5[1 + 7l_v(z)]\rho_{air}v_m^2(z) \qquad\qquad\text{(G)}$$

Wind Pressure on Surface

The external pressure coefficient is calculated as follows.

$$W_e = q_p(Z_e).C_{pe}\ \text{-----(H)}$$

where

$q_p(Z_e)$ = peak velocity pressure
Z_e = reference height of the external pressure
C_{pe} = pressure coefficient of the external pressure

In this problem:

b = 36.375 m
d = 14.45 m
h = 23.6 m < b

As per clause 7.2.2, a building whose height h is less than b should be considered to be one part, so the reference height of the vertical walls Z_e is calculated as follows. As $h \le b$, as per figure 7.4 of EN 1991-1-4: 2010:

$$Z_e = h = 23.6 \text{ m}$$

Again, as per figure 7.5 of EN 1991-1-4: 2010:

e = b or 2h, whichever is smaller; here b = 36.375 m and 2h = 47.2 m, so e = 36.375 m

As e > d (d = 14.45m), the side elevation of the side wall will be as per Figure 5.5. Thus:

A = e/5 = 7.3 m
B = d–e/5 = 7.15 m
h/d = 23.6/14.45 = 1.63 m
(Refer Figure 7.5 of EN 1991.1.4 2010)

As per Figure 5.5, the windward side is D, the leeward side E and the side walls are A, B.

As per figure 7.1 of EN 1991-1-4: 2010, the external pressure coefficient for the vertical wall, C_{pe}, has been computed in Table 5.9.

The internal pressure coefficient is calculated as follows.

As per clause 7.2.9 (6), note 2, of EN 1991-1-4: 2010, for a building without a dominant face, $c_{pi} = +0.2$ and -0.3. The internal pressure:

$$W_i = q_p(z_e).c_{pi} \text{ ---(I)}$$
where

Wi = internal wind pressure, Pa
$q_p(z)$ = peak pressure, Pa
c_{pi} = internal pressure coefficient

The wind pressure has been calculated for the wall, and the results are furnished in Table 5.10.

Seismic Analysis

The seismic parameters are considered as per BSEN 1998-1: 2004. Input is provided to the STAAD Pro CE software under the "Seismic" definition.

An elastic response of Type I is allowed for
As per table 3.1, clause 3.1.2, of BSEN 1998-1: 2004, the ground type is taken
     to be "C"
As per clause 3.2.1, the design ground acceleration $a_g = \gamma_1 a_{gR}$

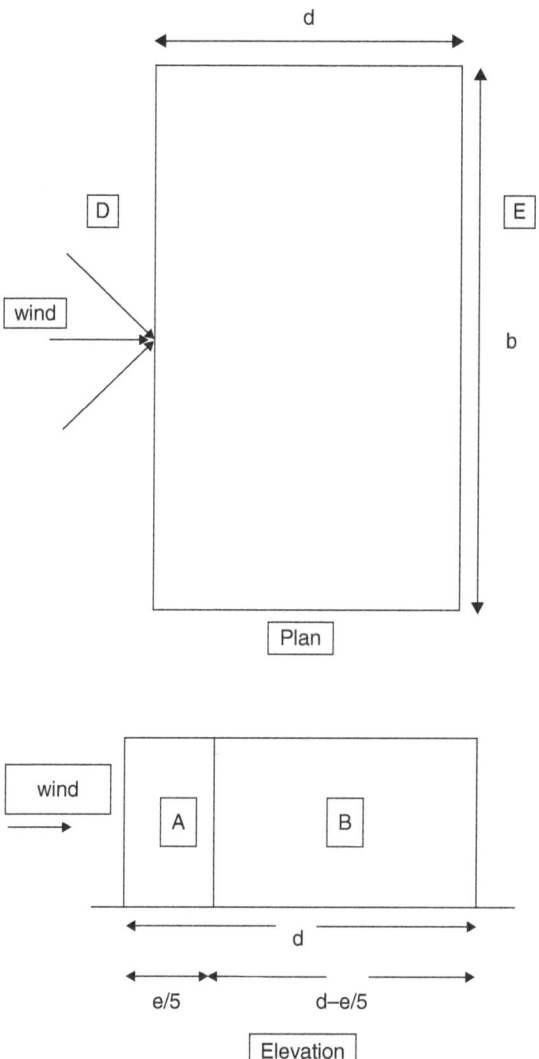

FIGURE 5.5 Wind load acting on a vertical wall (see figure 7.5 of EN 1991.1.4: 2010)

TABLE 5.9
External wind pressure coefficients

h/d	A	B	C	D	E
5	−1.2	−0.8		+0.8	−0.7
1.63	**−1.2**	**−0.8**		**+0.8**	**−0.469**
1	−1.2	−0.8		+0.8	−0.5

TABLE 5.10
Wind pressure on wall

Zone	w_e		w_i		Combined $w_{e \text{ and }} w_i$	
	–cpe	+cpe	+cpi	–cpi	Minimum value	Maximum value
A	–1.37		0.23	–0.34	–1.60	–1.03
B	–0.91				–1.14	–0.57
D		0.91			0.68	1.25
E	–0.54				–0.77	–0.20

where

γ_1 = importance factor
a_{gR} = peak ground acceleration
From Table 4.3, the importance class of the building is II
As per clause 4.2.5 5(P) of BSEN 1998-1: 2004, $\gamma_1 = 1.0$

Reference is made here to the peak ground acceleration proposed by Ambraseys, Simpson and Bommer (1996) for intra-plate seismicity in Europe. The attenuation of a_g is given by the expression:

$$\log(a_g) = -1.48 + 0.27M - 92\log(R)$$

where

M = the magnitude of variation of the ground motion
R = the epicentral distance

Considering, M = 6.5 and R = 30 km:

$$a_g = 0.23(g)$$

From Table 3.2, S = 1.15, $T_B(S) = 0.2$, $T_C(S) = 0.6$ and $T_D(S) = 2.0$. This elastic response curve is generated by the software.

Damping = 5%

Taking the building model as a frame system (refer to clause 5.2.2.1 1(P)) and a concrete building with medium ductility DCM (clause 5.2.1 4(P)),

The behavior factor q = $q_0.k_w$

where

q_0 = the basic behavior factor
k_w = a factor reflecting the prevailing failure mode in structural systems with walls

The basic value of the behavior factor as per Table 5.1:

$$q_0 = 3.0 \ (\alpha_u/\alpha_1)$$

and clause 5.2.2.2 5(P):

$$(\alpha_u/\alpha_1) = 1.3$$

As per clause 5.2.2.2 11(P), k_w = 1.0. Therefore:

$$Q = 3(1.3)(1.0) = 3.9$$

As per EN 1998-1, clause 4.4.2.3 3(P), in multistoried buildings the formation of the soft story plastic mechanism need to be prevented. In frame building with two or more stories, the following condition is to be satisfied at all joints of primary and secondary seismic beams with primary seismic columns:

$$\sum M_{Rc} \geq 1.3 \sum M_{Rb}$$

where

$\sum M_{Rc}$ = the sum of the design values of the moments of resistance of the columns framing the joint
$\sum M_{Rb}$ = the sum of the design values of the moments of resistance of the beams framing the joint

All the above-mentioned data have been provided to the STAAD Pro CE software and an analysis could be carried out.

5.5   COMPARISON OF BASIC PARAMETERS STIPULATED IN INDIAN, AMERICAN AND BRITISH CODES

A comparison of the basic parameters stipulated in Indian, American and British codes is made in Table 5.11.

TABLE 5.11

Comparison of basic parameters of wind and seismic load assessment

Parameters	Indian code	American code	British code
Wind force calculation			
	IS 875 (Part 3), 2015	ASCE 7–10	EN 1991.1.4: 2010
Basic wind speed	V_b; represented in figure 1 of the code	V; represented in figures 26.5-1A, B and C of ASCE 7–10	$V_b = C_{dir}C_{season}V_{b,0}$ where V_b = basic wind velocity in m/s C_{dir} = directional factor C_{season} = seasonal factor $V_{b,0}$ = fundamental value of the basic wind velocity (DIN National Annex for EN 1991-1-4)
Design wind pressure	$p_d = k_d.k_a.k_c.p_z$ (N/m²) where $p_z = 0.6.v_z^2$ (N/m²) where $v_z = k_1.k_2.k_3.k_4.V_b$ where k_1 to k_4 are different factors depending upon risk, terrain height, topography, cyclonic region respectively. k_d, k_a and k_c are different factors, namely directionality factor, area averaging factor and combination factor respectively.	Q = velocity pressure, in psf, given by the formula in clause 29.3.2 of ASCE 7–10: $q = 0.613K_zK_{zt}K_dV^2$ (N/m2) where K_z = velocity pressure coefficient K_{zt} = topographic factor K_d = wind directionality factor V = basic wind speed, in m/s	$q_b = 0.5\rho_{air}V_b^2$ where q_b = design wind pressure, in Pa ρ_{air} = density of air (1.25 kg/cu.m.) V_b = basic wind velocity, in m/s $q_p(z) = 0.5[1 + 7Iv(z)]\rho_{air}v_m^2(z)$ q_p = peak velocity pressure
Wind load on building	For building as a whole: $F = C_f.A_e.p_d$ where C_f = force coefficient A_e = effective frontal area of building	p = design wind pressure $p = q(GC_p)-qi(GC_{pi})$ where G = gust effect factor C_p = external pressure coefficient	$W_e = q_p(Z_e).C_{pe}$ where Z_e = reference height of the external pressure C_{pe} = pressure coefficient of external pressure

TABLE 5.11   (Continued)
Comparison of basic parameters of wind and seismic load assessment

Parameters	Indian code	American code	British code
	For individual structural elements of building: $F = (C_{pe} - C_{pi}).A.p_d$ where C_{pe} = external pressure coefficient C_{pi} = internal pressure coefficient	C_{pi} = internal pressure coefficient C_p values are different for windward, leeward and side walls. $F = A_e\, p$ A_e = effective area of building subjected to wind load	$W_i = q_p(z_e).c_{pi}$ c_{pi} = internal pressure coefficient. Sum of W_e and W_i will give net wind pressure. $F = A_e$. Net pressure A_e = effective area of building subjected to wind load

Seismic analysis

	IS 1893 (Part 1), 2016	ASCE 7–10, IBC 2012	BSEN 1998-1: 2004
Zone	As per figure 1 of IS 1893 (Part 1), 2016, India is divided into four earthquake zones, namely zone II to zone V	Risk-adjusted maximum considered earthquake (MCE_R) ground motion parameters S_S and S_1 are given in figures 22.1 to 22.6. S_S is the risk-adjusted MCE_R, 5%-damped, spectral response acceleration parameter at short periods. S_1 is the mapped MCE_R ground motion, 5%-damped, spectral response acceleration parameter at a period of 1 sec	Depending on the local hazards, the national authorities have created different zones. The hazard is described in terms of a single parameter – i.e., the value of the reference peak ground acceleration on type A ground: a_{gR}. Additional parameters required for specific types of structures are given in the relevant parts of EN1998-1: 2004.
Site classification as per soil condition of site	Site are classified as follows: rocky or hard soil sites medium stiff soil sites soft soil sites	Depending on the soil character, sites are classified in six categories: A to F	Ground classification goes from A to E type, depending on the value of the average shear wave velocity, $V_{S,30}$, and SPT (N_{SPT}), which is the number of blows per 30 cm penetration.

TABLE 5.11   (Continued)
Comparison of basic parameters of wind and seismic load assessment

Parameters	Indian code	American code	British code
Importance factor	1.5 for critical lifeline structure 1.2 for business continuity structure 1.0 for all other structures	Depending on the risk category, I to IV, different importance factors are considered in ASCE 7–10.	Building structures are subdivided into four importance classes, I to IV, as per increasing importance of the structure. Importance factors for these four types of building structure are 0.8, 1.0, 1.2 and 1.4 respectively.
Response reduction/ response modification factor	For an Reinforced Concrete building with an ordinary moment-resisting frame (OMRF): 3.0 For an Reinforced Concrete building with a special moment-resisting frame (SMRF): 5.0 For a steel building with an OMRF: 3.0 For a steel building with an SMRF: 5.0	As per table 12.2-1 Ordinary reinforced concrete moment frame: 3.0 Intermediate reinforced concrete moment frame: 5.0 Special reinforced concrete moment frame: 8.0 Steel ordinary moment frame: 3.5 Steel intermediate moment frame: 4.5 Steel special moment frame: 7.0	Behaviour factor is considered which is similar to Response reduction factor of other codes
Approximate natural period	Different formulae have been suggested for different types of frame, such as: Bare moment-resisting frame buildings, buildings with Reinforced Concrete structural walls, and other buildings. For all other buildings: $T_a = 0.09h/\sqrt{d}$ where h = height of the building, in m d = base dimension at plinth level in the direction of the earthquake shaking, m	Here, $T_a = C_t h_n^x$ where h_n = structure height, in m C_t and x are different for different moment-resisting frames, such as steel moment-resisting frames, concrete moment-resisting frames, etc. For all other structural systems: $T_a = 0.0488\, h_n^{0.75}$	For buildings up to 40 m height: fundamental period of vibration $T_1 = C_t H^{3/4}$ where H = height of building, in m $C_t = 0.085$ for moment-resisting space steel frames, 0.075 for moment-resisting space concrete frames and for eccentrically braced steel frames, 0.050 for all other structures

TABLE 5.11   (Continued)
Comparison of basic parameters of wind and seismic load assessment

Parameters	Indian code	American code	British code
Base shear	$V_b = A_h W$	$V = C_s W$	$F_b = S_d(T_1).m.\lambda$
	where	where	where
	A_h = design horizontal acceleration coefficient W = seismic weight of the building	Cs = seismic response coefficient W = effective seismic weight	$S_d(T_1)$ = ordinate of spectrum at period T_1 M = total mass of the building λ = correction factor

Bibliography

1   ACI 318-08 2008 *Building Code Requirements for Structural Concrete and Commentary.* Farmington Hills, MI: American Concrete Institute.

2   ACI 318 2014 *Building Code Requirements for Structural Concrete.* Farmington Hills, MI: American Concrete Institute.

3   Ambraseys, N.N., Simpson, K.A., and Bommer, J.J. 1996 Prediction of Horizontal Response Spectra in Europe. *Earthquake Engineering Structural Dynamics*, 25(4), 371–400.

3   Arnold, C., and Reitherman, R. 1982 *Building Configuration and Seismic Design: The Architecture of Earthquake Resistance.* New York: John Wiley.

4   ASCE/SEI 7–10 2013 *Minimum Design Loads for Buildings and Other Structures.* Reston, VA: American Society of Civil Engineers.

5   BS 8110–1 1997 *Structural Use of Concrete*, part 1, *Code of Practice for Design and Construction.* London: British Standards Institution.

6   Chopra, A.K. 1982 *Dynamics of Structures: A Primer.* Oakland, CA: Earthquake Engineering Research Institute.

7   Chopra, A.K. 2012 *Dynamics of Structures: Theory and Application to Earthquake Engineering*, 4th edn. Upper Saddle River, NJ: Prentice Hall.

8   Cook, N.J. 1985 *The Designer's Guide to Wind Loading of Building Structures*, part 1. London: Butterworths.

9   Davenport, A.G. 1967 Gust Loading Factors. *Journal of the Structural Division*, 93(3), 11–34.

10   Dowrick, D.J. 1987 *Earthquake Resistant Design for Engineers and Architects*, 2nd edn. New York: John Wiley.

11   EN 1990: 2002 + A1: 2005 2010 *Eurocode: Basis of Structural Design.* Brussels: European Committee for Standardization.

12   EN 1991-1-1 2002 *Eurocode 1: Actions on Structures*, part 1-1, *General Actions: Densities, Self-Weight, Imposed Loads for Buildings.* Brussels: European Committee for Standardization.

13   EN 1991-1-4: 2005 + A1: 2010 2010 *Eurocode 1: Actions on Structures*, part 1-4, *General Actions: Wind Actions.* Brussels: European Committee for Standardization.

14   EN 1998-1 2004 *Eurocode 8: Design of Structures for Earthquake Resistance*, part 1, *General Rules, Seismic Actions and Rules for Buildings.* Brussels: European Committee for Standardization.

15   Holmes, J.D. 1985 Recent developments in the codification of wind loads on low-rise structures. In *Proceedings of the Asia–Pacific Symposium on Wind Engineering, Roorkee, India, December 1985*, iii–xvi. Roorkee: University of Roorkee.

16   IBC2012 2011 *2012 Building Code.* Washington, DC: International Code Council.

17   IITK-GSDMA-EQ05-V4.0 2005 *Proposed Draft Provisions and Commentary on Indian Seismic Code IS:1893 (Part 1).* Kanpur: Indian Institute of Technology Kanpur and Gujarat State Disaster Mitigation Authority.

18   IITK-GSDMA-Wind02-V5.0 2015 *IS:875 (Part 3): Wind Loads on Buildings and Structures: Proposed Draft and Commentary.* Kanpur: Indian Institute of Technology Kanpur and Gujarat State Disaster Mitigation Authority.

19   IS 1080,1985 (reaffirmed 2002) 2002 *Code of Practice for Design and Construction of Shallow Foundations in Soils (Other than Raft, Ring and Shell)*, 2nd rev. New Delhi: Bureau of Indian Standards.

20   IS 13920, 2016 2016 *Ductile Design and Detailing of Reinforced Concrete Structures Subjected to Seismic Forces: Code of Practice.* New Delhi: Bureau of Indian Standards.

21   IS 15498, 2004 (reaffirmed 2020) 2020 *Guidelines for Improving the Cyclonic Resistance of Low Rise Houses and Other Buildings/Structures.* New Delhi: Bureau of Indian Standards.

22   IS 1893 (Part 1), 2016 2016 *Criteria for Earthquake Resistant Design of Structures*, part 1, *General Provisions and Buildings*, 6th rev. New Delhi: Bureau of Indian Standards.

23   IS 1904, 1998 (reaffirmed 2006) 2006 *Code of Practice for Design and Construction of Foundations in Soils: General Requirements.* New Delhi: Bureau of Indian Standards.

24   IS 2911 (Part 1 to 4), 2010 2010 *Code of Practice for Design and Construction of Pile Foundations.* New Delhi: Bureau of Indian Standards.

25   IS 2950 (Part 1), 1981 (reaffirmed 2008) 2008 *Code of Practice for Design and Construction of Raft.* New Delhi: Bureau of Indian Standards.

26   IS 43262013 2013 *Earthquake Resistant Design and Construction of Buildings: Code of Practice.* New Delhi: Bureau of Indian Standards.

27   IS 456, 2000 2000 *Plain and Reinforced Concrete: Code of Practice*, 4th rev. New Delhi: Bureau of Indian Standards.

28   IS 875 (Part 1), 1987 (reaffirmed 2008) 2008 *Design Loads (Other than Earthquake) for Buildings and Structures*, part 1, *Dead Loads: Unit Weight of Building Materials and Stored Materials.* New Delhi: Bureau of Indian Standards.

29   IS 875 (Part 2), 1987 (reaffirmed 2008) 2008 *Design Loads (Other than Earthquake) for Buildings and Structures*, part 2, *Imposed Loads.* New Delhi: Bureau of Indian Standards.

30   IS 875 (Part 3), 2015 2015 *Design Loads (Other than Earthquake) for Buildings and Structures: Code of Practice*, part 3, *Wind Loads*, 3rd rev. New Delhi: Bureau of Indian Standards.

31   IS 875 (Part 4), 1987 (reaffirmed 1997) 1997 *Design Loads (Other than Earthquake) for Buildings and Structures*, part 4, *Snow Loads.* New Delhi: Bureau of Indian Standards.

32   IS 875 (Part 5), 1987 (reaffirmed 1997) 1997 *Design Loads (Other than Earthquake) for Buildings and Structures*, part 5, *Special Loads and Combinations.* New Delhi: Bureau of Indian Standards.

33   Joint Research Centre 2011 *Eurocode 8: Seismic Design of Buildings: Worked Examples.* Luxembourg: Publications Office of the European Union.

34   Melbourne, W.H. 1977 Cross-Wind Response of Structures to Wind Action. In *Proceedings of the Fourth International Conference on Wind Effects on Buildings and Structures: Heathrow 1975*, 343–358. Cambridge: Cambridge University Press.

35   Murty, C.V.R. 2005 *Earthquake Tips: Learning Earthquake Design and Construction.* Kanpur: Indian Institute of Technology Kanpur.

36   Park, R., and Paulay, T. 1975 *Reinforced Concrete Structures.* New York: John Wiley.

37   Paterson, D.A., and Holmes, J.D. 1993 Computation of Wind Flow over Topography. *Journal of Wind Engineering and Industrial Aerodynamics*, 46/47, 471–476.

38   Reynolds, C.E., Steedman, J.C., and Threlfall, A.J. 2008 *Reynolds's Reinforced Concrete Designer's Handbook*, 11th edn. Abingdon: Taylor & Francis.

39   Robertson, A.P., Paulay, T., and Priestley, M.J.N. 1992 *Seismic Design of Reinforced Concrete and Masonry Buildings.* New York: John Wiley.

40   Sachs, P. 1978 *Wind Forces in Engineering*, 2nd rev. edn. Oxford: Pergamon Press.

41   Shanmugasundaram, J., Annamalai, G., and Venkateswaralu, B. 1989 Probabilistic Models for Cyclonic Wind Speeds in India. In *Proceedings of the Second*

*Asia–Pacific Symposium on Wind Engineering, Beijing, China, June 26–29, 1989,*123–130. Oxford: Pergamon Press.

42   Saunders, J.W., and Melbourne, W.H. 1977 Tall Rectangular Building Response to Cross-Wind Excitation. In *Proceedings of the Fourth International Conference on Wind Effects on Buildings and Structures: Heathrow 1975*, 369–379. Cambridge: Cambridge University Press.

43   Simiu, E., and Scanlan, R.H. 1996 *Wind Effects on Structures: Fundamentals and Applications to Design*, 3rd edn. New York: John Wiley.

44   SP16, 1980 1980 *Design Aids for Reinforced Concrete to IS:456-1978*. New Delhi: Bureau of Indian Standards.

45   STAAD Pro CE 3D Analysis and Design Software. Exton, PA: Bentley.

46   STAAD Foundation Advance CE Comprehensive Foundation Design Software. Exton, PA: Bentley.

47   Taranath, B.S. 2010 *Reinforced Concrete Design of Tall Buildings*. Boca Raton, FL: CRC Press.

Index

Across wind response 17, 27, 193
Along wind effect 185
Acceleration responsespectrum 34
Aerodynamic roughness 27, 187
Along wind response 25
Architectural requirements 4
Area averaging factor 22, 73, 178, 220
Angle of incidence 79, 181
Aspect ratio 5, 8, 20
Axial forces 11, 58

Background factor 26, 185, 188, 189
Base isolation 29
Base shear 4, 11, 49, 51, 52, 53, 86, 197, 225
Beam-column joints 10, 12, 13, 33
Bending moment 11, 12, 13, 18, 55, 58, 67
Body wave 31
Braced frame 10

Cantilever method 58, 116–126
Capacity design 2, 10, 11, 29, 33
Center of mass 2, 7, 8, 52, 87, 194
Center of stiffness 2, 7, 8
Code 19, 58
Combination factor 22, 73, 179, 222
Column 3, 4, 7, 10, 11, 12, 13, 16, 18, 33
Confinement 3, 4, 11, 51
Collapse pattern 1, 12
Cutouts 1, 7, 12, 57

Drag Coefficient 24, 75, 180
Design criteria 32
Damping coefficient 26, 27, 185, 190, 194
Damping factor 51
Damping percentage 34
Dead load 17, 62
Design eccentricity 8
Design horizontal earthquake acceleration
    coefficient 54
Design hourly mean wind speed 25, 26, 27, 184,
    185, 186, 192
Design response 14
Design wind pressure 17, 21, 22, 23, 24, 72, 73,
    77, 78, 79, 179, 181, 209, 211, 213, 217, 222
Design wind speed 21, 72, 177
Detailing of reinforcements 50, 146, 167
Drag coefficient 22, 24, 73, 75, 77, 78, 180
Diaphragm 1, 6, 8, 9, 10, 11, 54, 87, 195
Dynamic response 14, 17
Ductility 1, 2, 3, 4, 10, 11, 13, 32, 50, 51, 55,
    164, 220

Earthquake load 16, 58
Earthquake forces 9, 11
Earthquake ground motion 34
Earthquake resistant design 32, 50, 58
Eccentricity 8, 15, 154, 155, 160, 166, 167, 194
Effective moment of inertia 16
Effective reduced frequency 26, 185, 191
Energy dissipation 2, 9, 10, 12, 13, 35
Epicenter 50
Equivalent static method 52, 81, 197
External pressure 23, 24, 181, 182, 209, 211
External pressure coefficient 23, 24, 181,
    182, 209

Fault 29, 30, 31
Frame analysis 116
Frontal area 24, 25, 73, 74, 76, 77, 78, 180, 185,
    192, 193, 222
Fixed end moment 92, 93, 94, 104, 105, 106
Force coefficient 24, 25, 180, 185, 192,
    193, 222
Foundation 5, 10, 11, 32, 33, 58

Gross moment of inertia 16
Ground motion 17, 31, 34, 51, 52, 220, 223
Gust factor 22, 25, 184, 185, 186, 191, 192

Height factor 26, 185, 188, 189
Horizontal irregularities 6
Hourly mean wind speed factor 26, 187
Hysteresis 1, 2, 13

Infill frames 12
Influence area 64, 65, 73, 74
Importance factor 21, 49, 53, 83, 212, 213,
    220, 224
Infill 12, 13, 56
Internal pressure 19, 23, 24, 79, 181, 182, 183,
    209, 210, 218, 223
Internal pressure coefficient 23, 24, 79, 180, 181,
    209, 210, 218, 223
Intensity 17, 19, 25, 26, 30, 31, 33, 84, 85, 90,
    91, 92, 103, 104, 185, 186, 188, 189, 190,
    194, 217
Irregularities 5, 6, 12, 16, 61

Lateral force 11, 54, 86, 87, 116
Lateral load resisting system 9, 51
Lateral stiffness 18, 32, 51, 54, 55, 56, 87
Live load 17, 52, 58, 54, 67, 84, 85, 116, 131, 175,
    177, 209

Liquefaction potential 34, 36, 37, 40, 44
Load combination 7, 141, 142, 143, 144, 145, 154, 165
Load factor 141
Load path 5, 8, 32
Lumped mass 54

Mass participation factor 197
Mode shape 14, 15, 27, 194, 195
Modal mass 54
Modal participation factor 5
Moment resisting frame 9, 10, 11, 12, 13, 50, 51, 83, 194, 222
MSK intensity scale 30

Natural frequency 16, 17, 20, 25, 26, 27, 184, 186, 189, 194
Node 11, 15, 73, 74, 75, 76, 77
Non-parallel system 8

One way shear 165, 167
Openings 6, 8, 19, 23, 32, 79, 80, 182
Ordinary moment resisting frame 10, 51, 224

P- Δ effect 9, 55
P-wave 30
Peak factor 25, 26, 27, 185, 186, 189, 190, 194
Peak ground acceleration 52, 53, 218, 221
Pile cap 54, 83, 87, 197
Plastic hinges 10, 33
Pressure coefficient 20, 22, 23, 24, 78, 79, 80, 181, 182, 183, 209, 210, 211, 212, 217, 218, 219, 222, 223
Pounding 14
Push over analysis 4, 55

Redundancy 1, 2, 4, 10, 12, 50, 51
Re-entrant corner 5, 6, 8
Response reduction factor 49, 50, 51, 52, 53, 83, 224
Response spectrum 34, 54, 197
Response Spectrum Method 54
Resonant response 26, 186
Roughness factor 25, 185, 188, 216

Strength 2
Serviceability 2
Seismicity 29, 220
Seismograph 31, 32
Seismic moment 31

Seismic weight 14, 34, 52, 53, 54, 63, 84, 86, 225
Seismic zone factor 36, 40, 42, 44, 49, 83
Seismic coefficient 52
Shear reinforcement 146, 148, 150
Shear wall 9, 11, 12
Slab 6, 13, 61, 66– 71
Slenderness ratio 5
Soft story 195, 221
Special moment resisting frame 10, 83, 224
Size reduction factor 26, 186, 190
Spectrum of turbulence 26, 186, 191
Stiffness 1, 2, 3, 4, 6, 7, 8, 11, 12, 13, 15, 16, 17, 32, 33, 34, 51, 55, 87, 89, 90, 101
Site location 177
Story drift 11
Story shear 54, 55, 87, 195
Subsystems 9
Substitute frame 58, 89

Turbulence intensity 188
Terrain 21, 22, 26, 71, 178, 185, 186, 187, 188, 209, 215, 216, 222
Time period 5, 9, 16, 18, 19, 20, 213
Torsional effects 12
Topography 21, 22, 71, 178, 222
Turbulence intensity 19, 25, 26, 27, 185, 186, 188, 189, 190, 194, 217
Two-way shear 167

Unreinforced masonry infill 56

Vertical irregularities 6, 12
Vortex shedding 17, 20, 25, 185
Vortex shedding frequency 20

Weak storey 14
Wind pressure 16, 17, 19, 21, 22, 23, 25, 71, 72, 77, 78, 79, 80, 81, 177, 178, 179, 180, 181, 185, 192, 194, 209, 211, 212, 213, 217, 218, 219, 220, 222, 223
Working drawings 58, 169–174, 198–205
Wind turbulence 217
Wind angle 79, 80, 81, 182, 183, 184
Wind load 16, 17, 21, 23, 25, 27, 71, 73, 77, 78, 80, 128, 130, 180, 181–195

Yielding 4, 10, 11

Zone factor 36, 40, 42, 44, 49, 50, 52, 53, 83